D0331919

THE
MOON
BOOK

THE
MOON
BOOK

Fascinating Facts about the
Magnificent, Mysterious Moon

Kim Long

SCIENCE ADVISOR
Larry Sessions
FORMER DIRECTOR, NOBLE PLANETARIUM (FORT WORTH, TEXAS)
FORMER STAFF ASTRONOMER, GATES PLANETARIUM (DENVER, COLORADO)

Johnson Books
BOULDER

Copyright © 1998, 1988 by Kim Long

All rights reserved. No part of this publication may be reproduced or transmitted in any form or by any means, electronic or mechanical, including photocopy, recording, or any information storage and retrieval system, without permission in writing from the publisher.

Published in the United States by Johnson Books, a division of Johnson Publishing Company, 1880 South 57th Court, Boulder, Colorado 80301

9 8 7 6 5

Cover design: Debra B. Topping
Cover photograph: © Lick Observatory
Moon photographs on pages 27, 28, 68, 56, 120, 116:
    © Colleen Gino and Scott W. Teare
All illustrations by the author unless otherwise indicated.

Library of Congress Cataloging-in-Publication Data
Long, Kim.
    The moon book: fascinating facts about the magnificent,
    mysterious moon / Kim Long. — Rev. and expanded ed.
        p.   cm.
    Includes bibliographical references and index.
    ISBN 1-55566-230-7 (alk. paper)
    1. Moon — Handbooks, manuals, etc. I. title.
    QB581.L66 1998
    523.3—dc21   98-26266
                                                  CIP

Printed in the United States by
Johnson Printing
1880 South 57th Court
Boulder, Colorado 80301

CONTENTS

ACKNOWLEDGMENTS

Cathie and Tim Havens, S&S Optika (Englewood, Colorado)
Larry Sessions
Colleen Gino
Michael McNierney
Kathleen Cain
Roger Sinnott, *Sky & Telescope Magazine*
Guy Ottewell, Furman University (Greenville, South Carolina)
Brian and Thomas Bisque, Software Bisque (Golden, Colorado)
Alan Hale, Celestron International (Torrance, California)
The U.S. Naval Observatory
The Denver Public Library
Auraria Library, Metropolitan State College
Norlin Library, University of Colorado
Oliver C. Lester Library of Mathematics & Physics,
    University of Colorado
The Bloomsbury Review (Denver, Colorado)
The Tattered Cover Bookstore (Denver, Colorado)

Special thanks in absentia to Leroy Doggett of the U.S. Naval
Observatory for his initial help in the conception and research
of *The Moon Calendar* and the first edition of *The Moon Book*.

"Tacitae per amica silentia lunae." — VIRGIL (*Aeneid*)

INTRODUCTION

The first edition of *The Moon Book* was created in 1988 as a publishing venture to answer questions related to *The Moon Calendar*, another publishing adventure that this author has produced annually sine 1981. In 1988, twelve years had passed since the last unmanned lunar exploratory mission and sixteen years since humans had visited the Moon (Apollo 17). Analysis of lunar material returned from the surface was ongoing, revealing insight into the makeup and origin of the Moon. But at that point, emphasis seemed to have shifted away from further lunar exploration, a change due in part to the evolution of political and military goals and an economic focus on space activity closer to Earth, mostly space shuttle missions.

In the past few years, however, momentum seems to be building anew for more Moon-based exploration. Some of the new interest is linked to the apparent discovery of a source of water there, a major factor for future lunar usefulness. Aside from practical activities involved in the exploration of space, the Moon also is poised for a renaissance of interest from those bound to this planet. Along with astronomy in general, a large portion of the population — of all ages — has been newly inspired by the bountiful production of images from the Hubble Space Telescope and the recent visit by the Hale-Bopp comet.

Connections to information about the Moon and astronomy have also dramatically improved since the first edition of *The Moon Book*, primarily due to the birth and rapid spread of the World Wide Web and the proliferation of personal computers.

Meanwhile, the Earth's nearest neighbor continues its perpetual cycle, providing the clearest, most obvious, and most continuous educational object in the sky. This edition adds new illustrations and reference material as well as maintaining features already proven useful in satisfying curiosity about the Moon. Hopefully, an increasing number of people will look up, marvel, and ask new questions, perhaps requiring yet another new edition sometime in the future.

THE MOON'S ORBIT

The Moon orbits around the Earth in an elliptical path. An ellipse is an elongated circle; the degree to which it is elongated is called its eccentricity. The eccentricity of the Moon's orbit is very slight, too small to accurately depict in an illustration that would fit in this book. The difference between the shortest (perigee) and longest (apogee) distance from the Earth to the Moon, in fact, is only about the width of four Earths (see page 119).

Over time, the elliptical shape of the Moon's orbit shifts slightly, causing the shortest and farthest distance in any given year to change. In the 20th century, the greatest distance was 252,731 miles (406,712 km) on March 2, 1984, and the shortest distance was 221,451 miles (356,375 km) on January 4, 1912. In the 21st century, the greatest distance will be 252,728 miles (406,707 km) on March 14, 2002. The shortest distance will be 221,535 miles (356,509 km) on November 14, 2016. For most purposes, the distances from Earth in the orbit are averaged, with a mean perigee and apogee used to express the extremes.

The orbit of the Moon, however, is more complex than just that of

The combined masses of two bodies in a binary system have a common center of mass, known as the barycenter. The barycenter of the Earth and the Moon is located about 3,000 miles out from the center of the Earth, about 900 miles under the surface. The actual location changes constantly because the distance between the Earth and the Moon changes during the Moon's monthly orbit.

BARYCENTER

1

one body moving around another. The Moon and the Earth are actually orbiting around a common center of gravity called a barycenter that is created by the combination of the two masses. The barycenter of the Earth-Moon system is in constant motion because both bodies are in motion, one orbiting around the other, and both rotating around their axis.

The Earth, being much larger than the Moon, "pulls" the barycenter much closer to itself than to the Moon. The barycenter is located below the surface of the Earth, about three quarters out (3,000 miles) from the center of the planet. When the Moon is farthest away in its orbit, the barycenter is also at its extreme distance, only 869 miles

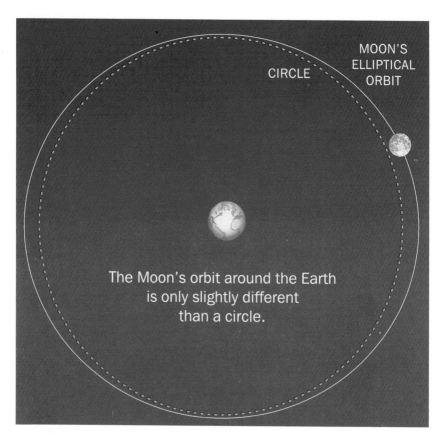

CIRCLE

MOON'S ELLIPTICAL ORBIT

The Moon's orbit around the Earth is only slightly different than a circle.

THE ROTATION OF THE MOON

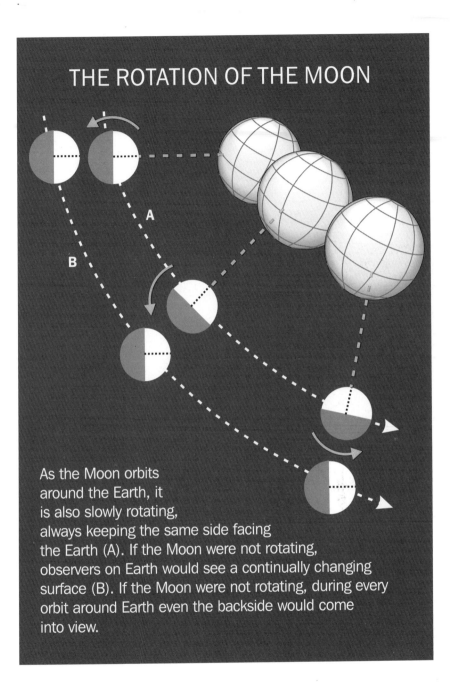

As the Moon orbits
around the Earth, it
is also slowly rotating,
always keeping the same side facing
the Earth (A). If the Moon were not rotating,
observers on Earth would see a continually changing
surface (B). If the Moon were not rotating, during every
orbit around Earth even the backside would come
into view.

(1400 km) below the Earth's surface. When the Moon is closest in its orbit, the barycenter also moves closer the center of the Earth, 1,180 miles (1900 km) below the surface. Because the Moon's orbit is an ellipse, the motion of the barycenter also forms an ellipse.

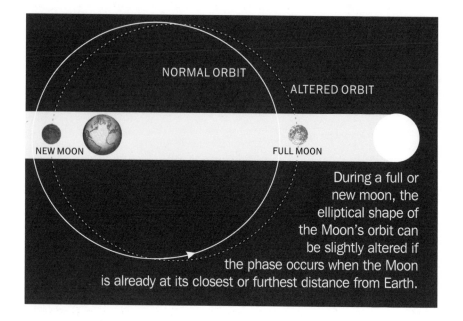

NORMAL ORBIT

ALTERED ORBIT

NEW MOON

FULL MOON

During a full or new moon, the elliptical shape of the Moon's orbit can be slightly altered if the phase occurs when the Moon is already at its closest or furthest distance from Earth.

MOON SPEED

Relative to the Earth, the Moon makes one rotation around its axis every 29½ days. This is the same time it takes for the Moon to complete one revolution around the Earth. The similarity of these two figures is no coincidence. The Moon rotated much faster in the past, but one of the effects of the Earth's gravity on the Moon over millions of years has been to slow down the rotation until the Moon has become "locked in step" with the Earth (see page 3).

The Earth rotates at about 1,000 miles an hour, as measured from a point on the surface at the equator. By comparison, the Moon rotates at only 10 miles an hour. On the other hand, the orbital speed of the Moon is much faster than its rotation. The average speed of the Moon on its monthly trip around the Earth is 2,287 miles an hour (3,683 kilometers an hour). However, the Moon's orbit is not a circle but an ellipse. The effect of an elliptical path on the speed of an orbiting object is to change the speed at different parts of the orbit.

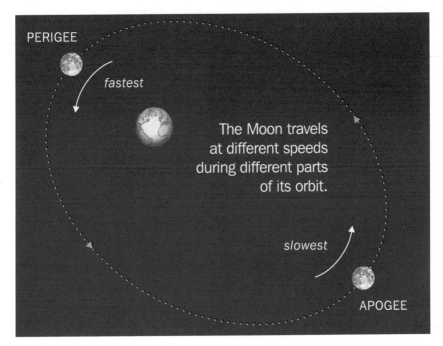

PERIGEE

fastest

The Moon travels at different speeds during different parts of its orbit.

slowest

APOGEE

When the Moon is closest to the Earth, it is traveling at its maximum speed, 2,470 miles an hour (3,978 km/hr). At the farthest point from the Earth, the speed is slowest, 2,173 miles an hour (3,499 km/hr). An observer on the surface of the Earth sees this orbital speed as the movement of the Moon across the sky at about one full Moon's width per hour, with the average distance being a little over 13 degrees a day (13°11', to be exact). At its theoretical fastest, the Moon could cover as much as 15°24' per day, an action that would require a combination of circumstances, primarily an extremely close perigee very close to the time of the full or new moon phase. At the other extreme, the Moon is less influenced by outside forces. Most lunar months, it slows down to just under 12 degrees of motion a day.

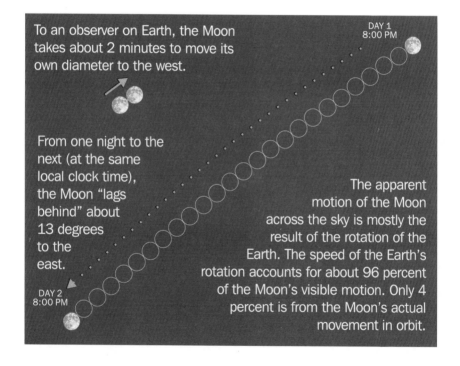

To an observer on Earth, the Moon takes about 2 minutes to move its own diameter to the west.

DAY 1
8:00 PM

From one night to the next (at the same local clock time), the Moon "lags behind" about 13 degrees to the east.

DAY 2
8:00 PM

The apparent motion of the Moon across the sky is mostly the result of the rotation of the Earth. The speed of the Earth's rotation accounts for about 96 percent of the Moon's visible motion. Only 4 percent is from the Moon's actual movement in orbit.

THE LUNAR MONTH

The Moon completes one orbit around the Earth in about 29½ days. This period is called a lunation, lunar month, or synodic month. A lunation begins at the exact time of a new moon and ends at the next new moon. One lunation is officially 29 days, 12 hours, 44 minutes, 2.8 seconds long. However, this measurement is an average, not a constant, and reflects monthly variations that occur over a long period of time (see also "Out of Phase," page 18).

If the lunar cycle is measured by timing the orbit in relationship to the position of a specific star, it is called a sidereal cycle. The sidereal lunar month is only 27 days, 7 hours, 43 minutes, 11.5 seconds long. The difference between the synodic month and the sidereal month is about 2 days per calendar month, which is about how much a full moon will "lag" behind the calendar, although the variation in length of a calendar month makes this an uneven rule of thumb.

A lunation is a very visible cycle that is easy to observe. Most people, however, notice the full moon instead of the new moon as it is more obvious, and think of the lunar cycle as running from full moon to full moon. The exception is traditional lunar calendars, most of which begin their months at the time of the new moon or the sighting of the first crescent moon.

Lunations are numbered in sequence. The sequence began with Lunation Number 1, designated by astronomers around the world as beginning with the new moon on January 16, 1923. There are 13 lunations in every calendar year because calendar months are longer than lunar months, with the exception of February.

THE ECLIPTIC

The Sun's path across the sky — actually caused by the revolution of the Earth around the Sun — is called the ecliptic. The ecliptic is kind of roadmap relative to the stars, creating a guide to the sky that can be used to find directions and tell time. Prominent stars and constellations are signs on this path, which also includes the twelve traditional constellations of the zodiac, used by both astronomers and astrologers, at least those in western cultures. Even though the stars are the same, other traditional and religious astrologies describe different constellations and zodiac groupings along the ecliptic (see page 91).

The Moon's path is tilted to the ecliptic by about 5 degrees, and since this tilt is itself rotating, the lunar path over time will eventually "sweep out" an area that is 5 degrees above and below the path of the Sun. Since the Moon is moving almost twelve times faster than the Sun around this stellar track, it makes a complete circuit in less than a month, a journey that takes the Sun one year to complete.

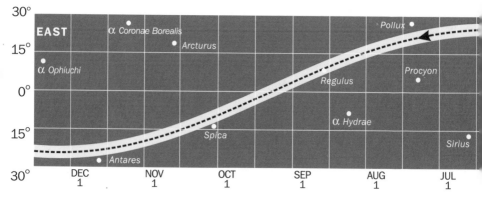

8

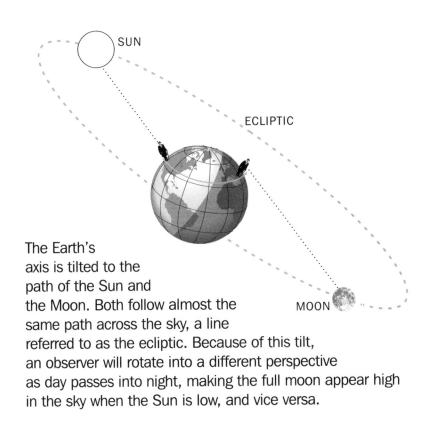

SUN

ECLIPTIC

MOON

The Earth's axis is tilted to the path of the Sun and the Moon. Both follow almost the same path across the sky, a line referred to as the ecliptic. Because of this tilt, an observer will rotate into a different perspective as day passes into night, making the full moon appear high in the sky when the Sun is low, and vice versa.

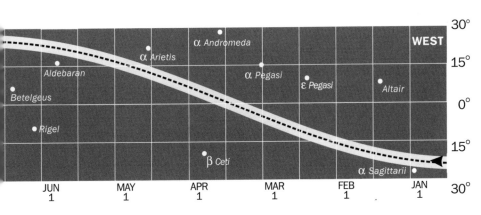

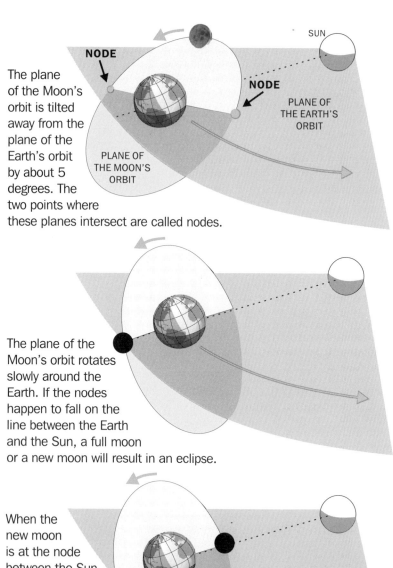

The plane of the Moon's orbit is tilted away from the plane of the Earth's orbit by about 5 degrees. The two points where these planes intersect are called nodes.

NODE

NODE

SUN

PLANE OF THE EARTH'S ORBIT

PLANE OF THE MOON'S ORBIT

The plane of the Moon's orbit rotates slowly around the Earth. If the nodes happen to fall on the line between the Earth and the Sun, a full moon or a new moon will result in an eclipse.

When the new moon is at the node between the Sun and the Earth, the result is a solar eclipse. A lunar eclipse results from a full moon at the opposite node.

LUNAR PHASES

The Moon appears to change shape as it moves through its monthly cycle. The Moon itself is not changing shape, but the part that is illuminated by the Sun does change, giving us the familiar and distinctive lunar phases. The four official phases that are included in calendars and almanacs are, in order, new moon, first quarter moon, full moon, and last quarter moon.

The new moon is actually invisible, as it is defined as an instant of time when the Moon is between the Sun and the Earth, and therefore lost in the bright light of the Sun. At this stage, not even a thin crescent is illuminated by the Sun and when the Moon is approaching and leaving the new moon phase, there is a period of two to four days when it cannot be easily seen, and a period of at least a day when it can't be seen at all.

The first and last quarter moons mark the halfway points between the new moon and full moon. The first quarter moon is always "first" and is distinguished by the illuminated half of the lunar surface being on the right-hand side. The last quarter moon is the reverse, with the illuminated half on the left-hand side.

The lighted part of the Moon always points the way to the Sun. If the lighted half is on the right, the Sun is on the right (west), meaning the Sun is ahead of the Moon. If the left half is lighted, the Moon is ahead of the Sun, and the Sun is on the left (east).

The old crescent moon shows light on the left and is seen in the east in early morning before the Sun rises.

The new crescent moon shows light on the right and is seen in the west in early evening after the Sun sets.

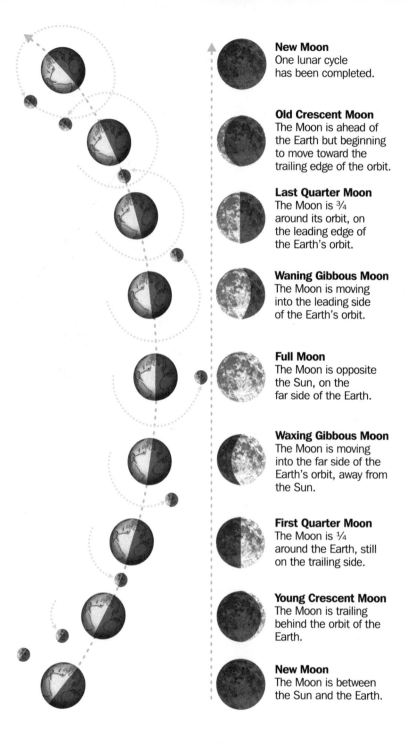

New Moon
One lunar cycle
has been completed.

Old Crescent Moon
The Moon is ahead of
the Earth but beginning
to move toward the
trailing edge of the orbit.

Last Quarter Moon
The Moon is ¾
around its orbit, on
the leading edge of
the Earth's orbit.

Waning Gibbous Moon
The Moon is moving
into the leading side
of the Earth's orbit.

Full Moon
The Moon is opposite
the Sun, on the
far side of the Earth.

Waxing Gibbous Moon
The Moon is moving
into the far side of the
Earth's orbit, away from
the Sun.

First Quarter Moon
The Moon is ¼
around the Earth, still
on the trailing side.

Young Crescent Moon
The Moon is trailing
behind the orbit of the
Earth.

New Moon
The Moon is between
the Sun and the Earth.

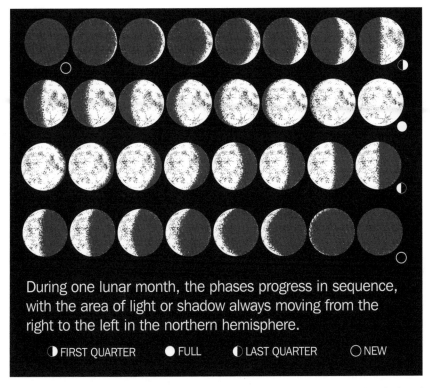

During one lunar month, the phases progress in sequence, with the area of light or shadow always moving from the right to the left in the northern hemisphere.

◑ FIRST QUARTER   ● FULL   ◐ LAST QUARTER   ○ NEW

The sequence of the lunar phases always proceeds with the lighted part of the Moon growing from right to left until the full moon, then receding from right to left until the new moon. You can always tell where the Moon is in its cycle if you remember that the changing pattern of light and dark always goes from right to left. If the Moon is light on the right side, the light will be expanding to the left; if the Moon is dark on the right side, the shadow will be expanding to the left.

The new moon is also called the "dark of the moon," as that is when it is totally dark. The period of darkness officially lasts for only a second — that point when the Moon is directly between the Sun and the Earth. However, observers on Earth don't see any "slivers" of light at the edges of the Moon just before and after this point because the new moon is too close to the Sun to see anything clearly.

During the period around the new moon, the Moon follows the Sun

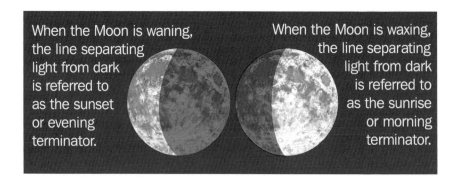

When the Moon is waning, the line separating light from dark is referred to as the sunset or evening terminator.

When the Moon is waxing, the line separating light from dark is referred to as the sunrise or morning terminator.

very closely and immediately after the new moon, it begins to fall behind the Sun. The first visible crescent moon (with the crescent on the right) is usually spotted two or three days after the new moon. Observers can see this young crescent moon, as it is sometimes called, just after sunset, with the Moon following the Sun down over the western horizon. The earliest that an observer on Earth has ever seen the young crescent moon is about 14 hours after new moon (see "First Sighting," page 43).

As the Moon "grows," or waxes, the lighted portion on the right gradually increases until it forms an almost perfect half circle, the first quarter moon. The quarter moons are created by the position of the Moon in relation to the Earth and Sun. At this point in the cycle, the orbit of the Moon has moved it to a position off to the side of the line connecting the Earth and the Sun. At the point of first quarter moon, the Moon is 90 degrees (perpendicular) to this imaginary line. Technically, this phase is called a quadrature.

The edge of light that gradually creeps across the lunar surface is called the terminator. The terminator is rarely an even line, as the rough surface and uneven curvature of the Moon distort it, a phenomenon visible through binoculars or telescopes. But at the point of first and last quarter phase, naked eye observations from Earth usually show a reasonably straight perpendicular line. Because librations affect exactly what portion of the front face of the Moon is facing the Earth (see "Librations," page 53), during the first and last quarter phase, the terminator's exact location may vary. That is, the central meridian, which marks the exact north-south line through the center of the front

COMPARING MOON PHASES

DAY	Phase	Description
0	**NEW MOON**	Rises and sets with the Sun,
1		lags a few hours behind the Sun
2		
3	WAXING CRESCENT	
4		
5		
6		
7.4	**FIRST QUARTER**	Above horizon ½ in day, ½ at night,
8		lags 8–10 hours behind the Sun
9		
10	WAXING GIBBOUS	
11		
12		
13		
14.8	**FULL MOON**	Rises about sunset, sets about sunrise,
15		precedes Sun by 8–10 hours
16		
17		
18	WANING GIBBOUS	
19		
20		
21		
22.1	**LAST QUARTER**	Above horizon ½ at night, ½ day,
23		precedes Sun by a few hours
24		
25		
26	WANING CRESCENT	
27		
28		
29.5	**NEW MOON**	

surface, may not line up with the terminator. At extremes, the terminator may appear up to 7° 45' on either side of the central meridian.

As the shadow line moves across the surface, it is moving at a steady speed of about half a degree an hour, or 12.2° per day. If the Moon were a flat, two-dimensional object, this shadow would appear to move across the surface at a regular pace. But because it is a sphere, the shadow line is not always visible from the same angle to observers on Earth; it appears to move fastest across the surface between the first quarter and last quarter moon — with the full moon in between. Before and after this time, the shadow appears to move more slowly because we see less of the rounded surface around the edges of a sphere.

Patient observers are able to see this motion with the help of binoculars or a telescope. With a little magnification, observers may also note shadows created by the Sun's rays striking elevated features. These shadows can also be observed slowly creeping across the surface. During the waxing period, these shadows will be cast to the west (left); during the waning period, they fall to the east (right).

As the Moon continues waxing, it continues to fall farther and farther behind the movement of the Sun. The period after the first quarter moon when the terminator continues to move to the right, lighting more and more of the surface, is still part of the waxing moon. It is also sometimes referred to as the gibbous moon (which can also refer to the period after the full moon up until the last quarter moon), or more specifically, the waxing gibbous moon. A gibbous moon exhibits a lighted "bulge"; it lasts from the first quarter moon up until the time of the full moon — but not including the full moon — and then again after the full moon up until the last quarter moon.

The Moon becomes full when the orbital path of the Moon carries it directly opposite of the Sun, on the other side of the Earth. At this position, it receives the direct light of the Sun across the full face, forming the distinctive full moon image. Another term for full moon is "moon in opposition," because it is opposite the Sun. Being opposite the Sun at the time of full moon, the Moon rises just as the Sun is setting.

The full moon immediately begins diminishing, even though it might not be visibly apparent for a day or two, because the spherical shape of the Moon can hide the first effects of the growing shadow on the right-hand side. The period after the full moon up until the new moon is called a waning moon. The period between the full moon and the last quarter moon is more specifically called a waning gibbous moon. During this part of the cycle, the Moon is catching up to the Sun. The sunlight that illuminates it comes from "behind," so the lighted surface starts to diminish from the right.

The growing shadow on the right-hand side of the face gradually increases until it forms an almost perfect half circle. At this point, the phase is called the last quarter moon. It is opposite to the first quarter moon, with the lighted half now being on the left side. Like the first quarter moon, this phase is also technically referred to as a quadrature, and the Moon is now positioned at a right angle (perpendicular) to the imaginary line connecting the Sun and the Earth.

The Moon continues to "shrink" after the last quarter, with the growing shadow obscuring more and more of the face. In less than a week, the entire face will be in shadow, and the cycle will be back at the beginning, the new moon. The last part of the visible lunar phase is often called the old crescent moon.

OUT OF PHASE

One lunar cycle, from new moon to new moon, is completed about every 29½ days. The "official" length of this period is 29.53059 days, or 29 days, 12 hours, 44 minutes, and 2.8 seconds. Many people think of this figure as a constant; a regular benchmark that marks out lunar rhythm over time. Unfortunately, like many things associated with the Moon's movement, the truth involves a lot of variation.

During its monthly orbit around Earth, the Moon's speed is affected by many factors, including the gravitational forces of the Earth and the Sun and regular and irregular variations in its own orbit. The major effect on the length of a lunation comes from the force of the Sun.

From one month to the next, a lunation — the technical name for the period from one new moon to the next new moon — can vary from a few minutes to several hours, with an average variation of about fourteen hours over 100 years (13.44 hours, to be exact). Roger Sinnott, an editor at *Sky & Telescope Magazine*, calculated the variation over a much longer period, 500 years (from 1600 to 2100) and found that the mean variation was about nine hours. From 1900 to 2100, the extremes range from about six hours less than the average to a little more than seven hours greater than the average, according to calculations made by Jean Meeus, a noted Belgian mathematician and expert on astro-

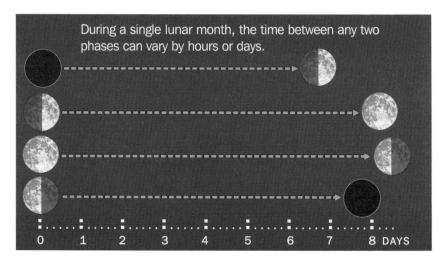

During a single lunar month, the time between any two phases can vary by hours or days.

0   1   2   3   4   5   6   7   8 DAYS

18

nomical cycles. The shortest lunation occurred on June 25 to July 24, 1903, a period of 29 days, 6 hours, and 35 minutes. The longest lunation was December 24, 1973, to January 23, 1974, a period of 29 days, 19 hours, and 55 minutes.

More confounding for lunar watchers is the fluctuation in the period of time between any two successive phases, full moon to last quarter moon, for example. This period may vary considerably from one phase to the next. If these periods were equal, each would span 7.38 days, or one quarter of the synodic month. But instead, they vary by as much as 11 percent from this standard — plus or minus 19 hours — making the length of time between two successive phases vary from six to nine calendar days, although most periods are not this extreme.

Phases and lunar months fluctuate, particularly compared to the daily calendars most people rely on, but there are patterns that appear over longer periods of time. Every 19 years, for example, a cycle of 235 lunar months (lunations) is completed and the new moon and full moons will fall on the same calendar days, with a few exceptions. In other words, in 1999, a new moon falls on January 17 and 19 years later in the year 2018, it also falls on January 17, a pattern known as the Metonic Cycle. It is subject to variation, particularly affected by location.

Another pattern was observed and named by early Greek astronomers. Based on an eight-year cycle, this phenomenon is called the *octaeteris*. Every eight years, the new moon and full moon fall on a calendar day that is one to two days later than the day in the first calendar year. In other words, a new moon falls on January 17 in 1999 and eight years later, in 2007, it falls two days later, on January 19. Moon phases also precede each other on almost the same dates every two years. The new moon in 1999, for example, falls on January 17 but two years later there is a last quarter moon one day earlier, on January 16 in 2001.

Another intriguing pattern is created by phases from one lunation to the next. The time of occurrence of a specific phase alternates from day to night. For example, if the exact time of a full moon falls during the day, in the next lunar cycle it will fall during the night. The same is true for each phase, alternating from day to night from one lunar month to another.

Another recurring cycle shows up about every 15 lunations. Given a particular phase, the same phase will occur within about 1½ hours after 15 lunations have passed, a period of a year and a few weeks or approximately every 433 days.

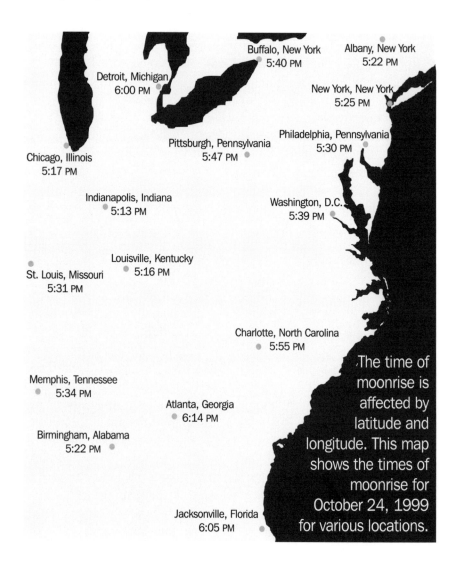

Buffalo, New York
5:40 PM

Albany, New York
5:22 PM

Detroit, Michigan
6:00 PM

New York, New York
5:25 PM

Pittsburgh, Pennsylvania
5:47 PM

Philadelphia, Pennsylvania
5:30 PM

Chicago, Illinois
5:17 PM

Indianapolis, Indiana
5:13 PM

Washington, D.C.
5:39 PM

St. Louis, Missouri
5:31 PM

Louisville, Kentucky
5:16 PM

Charlotte, North Carolina
5:55 PM

Memphis, Tennessee
5:34 PM

Atlanta, Georgia
6:14 PM

Birmingham, Alabama
5:22 PM

The time of moonrise is affected by latitude and longitude. This map shows the times of moonrise for October 24, 1999 for various locations.

Jacksonville, Florida
6:05 PM

20

MOONRISE AND MOONSET

The Moon rises and sets at different times every day because the calendar is based on a solar timetable, not a lunar one. On average, moonrise and moonset are about one hour later each succeeding day, but the times change considerably from one location to another. Both latitude and longitude have an effect on this change. In some northern latitudes (northern Canada, for instance), there is a more dramatic change from day to day in the times for moonrise and moonset.

During some periods of the lunar cycle, there are many days when the Moon

MOON DATA

To find local times for phases, rising, setting, position, and eclipses of the Moon, use the online calculator provided by the U.S. Naval Observatory:

http://riemann.usno.navy.mil/AA/data/

does not rise at all, or the opposite, does not set. During each lunar month, there is one day with no moonrise and one day with no moonset. This happens because the Moon "lags behind" the 24-hour day. The Moon actually has a 25-hour day (approximately). Therefore, for example, if the Moon sets at 11:50 P.M. on a Tuesday night, 25 hours later would be completely past Wednesday, and the next setting time would be about 12:40 A.M. on Thursday morning.

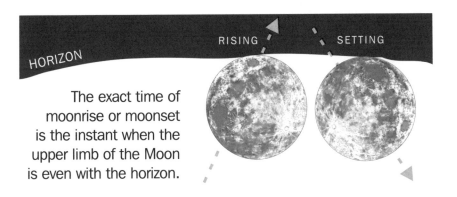

RISING   SETTING

HORIZON

The exact time of moonrise or moonset is the instant when the upper limb of the Moon is even with the horizon.

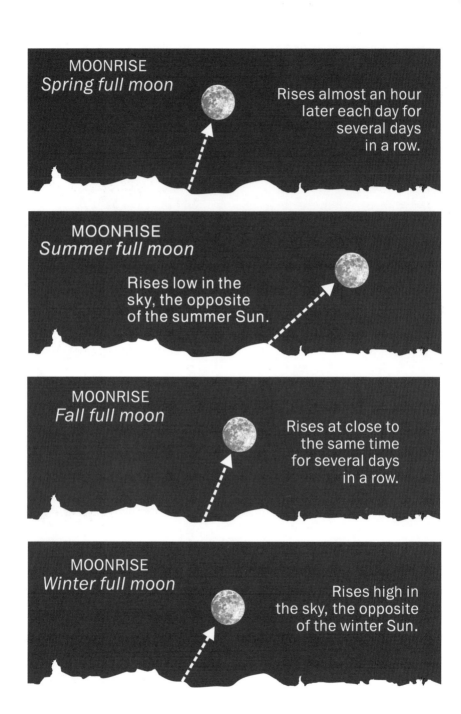

MOONRISE
Spring full moon

Rises almost an hour later each day for several days in a row.

MOONRISE
Summer full moon

Rises low in the sky, the opposite of the summer Sun.

MOONRISE
Fall full moon

Rises at close to the same time for several days in a row.

MOONRISE
Winter full moon

Rises high in the sky, the opposite of the winter Sun.

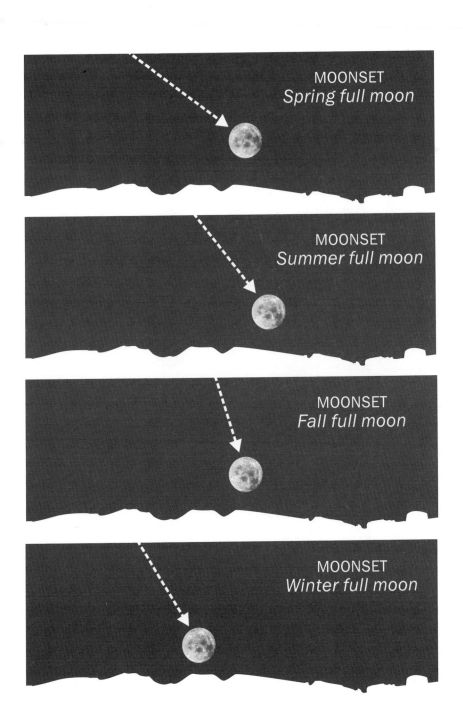

MOONSET
Spring full moon

MOONSET
Summer full moon

MOONSET
Fall full moon

MOONSET
Winter full moon

23

UNIVERSAL TIME FOR MERIDIAN OF GREENWICH

MOONRISE

Lat.	+40°	+42°	+44°	+46°	+48°	+50°	+52°	+54°	+56°	+58°	+60°	+62°	+64°	+66°
	h m	h m	h m	h m	h m	h m	h m	h m	h m	h m	h m	h m	h m	h m
Oct. 24	17 22	17 21	17 19	17 17	17 15	17 13	17 11	17 08	17 06	17 03	16 59	16 56	16 51	16 46
25	18 00	17 57	17 54	17 51	17 47	17 43	17 39	17 35	17 30	17 24	17 18	17 10	17 02	16 53
26	18 42	18 38	18 33	18 29	18 24	18 18	18 12	18 05	17 58	17 50	17 41	17 30	17 17	17 02
27	19 29	19 24	19 18	19 13	19 06	18 59	18 52	18 43	18 34	18 23	18 11	17 57	17 40	17 18
28	20 22	20 17	20 10	20 04	19 57	19 49	19 40	19 31	19 20	19 07	18 53	18 36	18 15	17 48
29	21 21	21 15	21 09	21 02	20 55	20 47	20 38	20 28	20 17	20 04	19 49	19 31	19 09	18 39
30	22 24	22 18	22 13	22 06	22 00	21 52	21 44	21 35	21 24	21 12	20 59	20 42	20 22	19 56
31	23 29	23 24	23 19	23 14	23 08	23 01	22 54	22 47	22 38	22 28	22 17	22 03	21 47	21 28
Nov. 1	..	..	..	..	..	..	..	..	23 54	23 46	23 38	23 28	23 16	23 03
2	0 33	0 30	0 26	0 21	0 17	0 12	0 07	0 01	..	..	..	..	..	..
3	1 37	1 34	1 32	1 29	1 26	1 22	1 18	1 14	1 10	1 05	0 59	0 53	0 45	0 36
4	2 39	2 38	2 36	2 35	2 33	2 31	2 29	2 27	2 25	2 22	2 19	2 16	2 12	2 07
5	3 40	3 39	3 39	3 39	3 39	3 39	3 38	3 38	3 38	3 38	3 37	3 37	3 36	3 36
6	4 39	4 40	4 41	4 42	4 44	4 45	4 46	4 48	4 50	4 52	4 54	4 57	4 59	5 03
7	5 38	5 40	5 43	5 45	5 48	5 50	5 54	5 57	6 01	6 05	6 10	6 15	6 22	6 29
8	6 36	6 40	6 43	6 47	6 51	6 55	7 00	7 05	7 11	7 17	7 25	7 33	7 43	7 55
9	7 34	7 38	7 43	7 47	7 53	7 58	8 05	8 11	8 19	8 28	8 38	8 50	9 03	9 20
10	8 30	8 35	8 41	8 46	8 53	8 59	9 07	9 15	9 25	9 36	9 48	10 03	10 21	10 43
11	9 24	9 30	9 36	9 42	9 50	9 57	10 06	10 15	10 26	10 38	10 53	11 10	11 31	11 59
12	10 16	10 22	10 28	10 35	10 42	10 50	10 59	11 09	11 21	11 34	11 49	12 08	12 31	13 02
13	11 04	11 09	11 16	11 22	11 30	11 38	11 47	11 57	12 08	12 21	12 36	12 54	13 16	13 46
14	11 48	11 53	11 59	12 05	12 12	12 19	12 27	12 37	12 47	12 59	13 12	13 28	13 48	14 12
15	12 28	12 32	12 37	12 43	12 49	12 55	13 02	13 10	13 19	13 29	13 40	13 53	14 09	14 28
16	13 04	13 08	13 12	13 16	13 21	13 26	13 32	13 38	13 45	13 53	14 02	14 12	14 24	14 37
17	13 38	13 41	13 44	13 47	13 51	13 54	13 58	14 03	14 08	14 13	14 19	14 26	14 34	14 44

MOONSET

Lat.	+40°	+42°	+44°	+46°	+48°	+50°	+52°	+54°	+56°	+58°	+60°	+62°	+64°	+66°
	h m	h m	h m	h m	h m	h m	h m	h m	h m	h m	h m	h m	h m	h m
Oct. 24	5 30	5 31	5 32	5 33	5 35	5 36	5 37	5 39	5 41	5 43	5 45	5 47	5 50	5 53
25	6 42	6 45	6 47	6 50	6 53	6 56	7 00	7 03	7 08	7 12	7 18	7 24	7 31	7 39
26	7 56	8 00	8 03	8 08	8 12	8 17	8 23	8 29	8 35	8 43	8 52	9 01	9 13	9 27
27	9 09	9 13	9 19	9 24	9 30	9 37	9 44	9 52	10 01	10 11	10 23	10 37	10 53	11 14
28	10 19	10 24	10 30	10 36	10 43	10 51	10 59	11 09	11 20	11 32	11 46	12 03	12 24	12 51
29	11 23	11 28	11 35	11 41	11 49	11 57	12 06	12 16	12 27	12 40	12 55	13 13	13 35	14 05
30	12 19	12 25	12 31	12 37	12 44	12 52	13 00	13 10	13 21	13 33	13 47	14 03	14 24	14 50
31	13 08	13 13	13 18	13 24	13 30	13 37	13 44	13 52	14 02	14 12	14 24	14 38	14 54	15 15
Nov. 1	13 50	13 54	13 58	14 03	14 08	14 13	14 19	14 26	14 33	14 41	14 50	15 01	15 13	15 28
2	14 26	14 29	14 33	14 36	14 40	14 44	14 48	14 53	14 58	15 04	15 10	15 18	15 26	15 36
3	14 59	15 01	15 03	15 05	15 07	15 10	15 12	15 15	15 18	15 22	15 26	15 30	15 35	15 41
4	15 29	15 30	15 30	15 31	15 32	15 33	15 34	15 35	15 36	15 38	15 39	15 41	15 43	15 45
5	15 58	15 57	15 57	15 56	15 56	15 55	15 55	15 54	15 53	15 52	15 51	15 50	15 48	15 47
6	16 26	16 25	16 23	16 21	16 19	16 17	16 15	16 13	16 10	16 07	16 04	16 01	15 56	15 52
7	16 55	16 53	16 50	16 47	16 44	16 41	16 37	16 33	16 28	16 23	16 18	16 11	16 04	15 56
8	17 27	17 23	17 19	17 15	17 11	17 06	17 01	16 55	16 49	16 41	16 33	16 24	16 13	16 01
9	18 01	17 56	17 51	17 46	17 41	17 35	17 28	17 21	17 12	17 03	16 53	16 40	16 26	16 09
10	18 38	18 33	18 27	18 21	18 15	18 08	18 00	17 51	17 41	17 30	17 17	17 02	16 44	16 22
11	19 20	19 14	19 08	19 01	18 54	18 46	18 37	18 28	18 17	18 04	17 50	17 32	17 11	16 43
12	20 06	20 00	19 53	19 47	19 39	19 31	19 22	19 12	19 00	18 47	18 32	18 13	17 50	17 19
13	20 56	20 50	20 44	20 38	20 30	20 22	20 14	20 04	19 53	19 40	19 25	19 07	18 45	18 16
14	21 51	21 45	21 40	21 34	21 27	21 20	21 12	21 03	20 53	20 42	20 28	20 13	19 53	19 29
15	22 48	22 44	22 39	22 34	22 28	22 22	22 15	22 08	22 00	21 50	21 39	21 27	21 12	20 53
16	23 49	23 45	23 42	23 38	23 33	23 29	23 23	23 18	23 11	23 04	22 56	22 47	22 36	22 23
17	..	..	..	..	..	..	..	..	..	..	..	..	..	23 56

.. .. indicates phenomenon will occur the next day.

Times of moonrise and moonset are listed in the *Astronomical Phenomena*, an annual publication produced by the U.S. Naval Observatory and available from the U.S. Government Printing Office.

MOON SIZE AND MOONLIGHT

The apparent size of the Moon to observers on Earth is about half of a degree. This is also the apparent size of the Sun. If you hold your hand up to the Moon with arm outstretched, this corresponds to about half the width of one finger. Sometimes, however, the Moon appears to be much larger. When rising and setting, for example, the Moon at full phase often appears to be larger than when it is directly overhead.

There are factors which can change the diameter and brightness of the Moon, but the rising and setting phenomena are actually optical illusions. At these times, closeness to the horizon adds a remarkable degree of visual influence. The eye is tricked into measuring the Moon against nearby objects, and the visual elements found on most horizons — buildings, trees, hills — create the impression of increased size.

This optical trick can be demonstrated by using cardboard masks to cut off the view of nearby objects. Make a mask with a viewing hole no larger than ¼ inch in diameter. Tape the mask to a yardstick so it can be kept at a fixed distance from your eye. As a full moon rises, view it through the mask when it just clears the horizon. Note the size of the Moon's disk compared to the size of the hole, then wait for the Moon to reach its zenith and view it again.

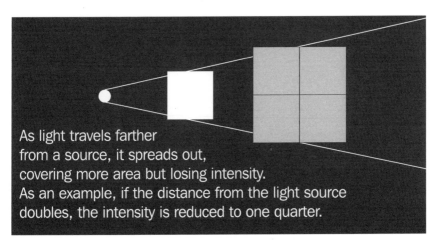

As light travels farther from a source, it spreads out, covering more area but losing intensity. As an example, if the distance from the light source doubles, the intensity is reduced to one quarter.

Although most of the apparent size change in the Moon is caused by visual trickery, there are physical effects which cause a measurable size difference, although not a large one. One of these effects is the difference in distance between an observer and the Moon from a horizontal perspective to a vertical one. This difference — the change from viewing the Moon on the horizon and at its zenith — is half the diameter of the Earth. However, this accounts for only a maximum difference of two percent in the size of the Moon, hardly enough to notice with the naked eye. And in any case, this size difference decreases the size of the Moon when it is at the horizon, not the opposite, because that is when it is farthest away from the observer.

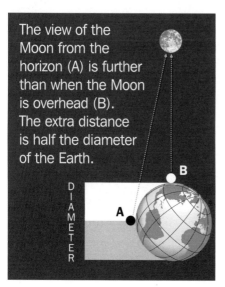

The view of the Moon from the horizon (A) is further than when the Moon is overhead (B).
The extra distance is half the diameter of the Earth.

A more significant difference is caused by the variation in the Moon's orbit around the Earth. This orbit is not round, but elliptical (oval) in shape, making some points in the orbit closer to the Earth than others. At the closest point (perigee), the Moon will appear measurably larger than when at the farthest point (apogee). The difference in diameter from one to the other is about 10 percent.

The Moon makes one orbit every 29.5 days, so there is one perigee and one apogee during every orbital period. However, the phases of the Moon are not synchronized with this cycle, and a full moon — the most visible phase — will typically occur on or near a perigee or apogee in only a few months in any calendar year. If the two cycles coincided in August, with the full moon occurring on about the same day as the perigee — give or take a few days — they might also coincide in September, with the date of perigee gradually "drifting" backwards through the calendar month faster than the date of full moon. By October or November, the date of perigee would be more than a few

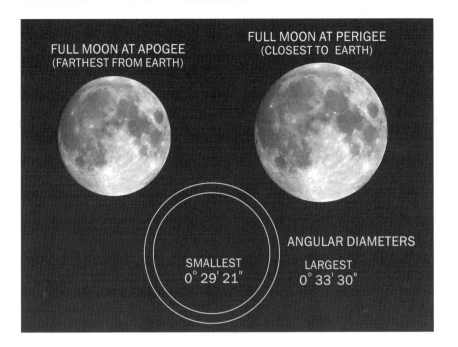

FULL MOON AT APOGEE
(FARTHEST FROM EARTH)

FULL MOON AT PERIGEE
(CLOSEST TO EARTH)

ANGULAR DIAMETERS

SMALLEST
0° 29' 21"

LARGEST
0° 33' 30"

days before the full moon, and the distance effect would thus be diminished.

The same kind of distance factors that change the size of the Moon can affect the amount of sunlight it reflects. Light is affected by the law of inverse squares — it decreases as the square of the distance. This happens because light spreads out as it gains distance; the farther from the light source, the more area that the same amount of light is covering. For example, if you double the distance between you and a lightbulb from 10 to 20 feet, the illumination would drop by 75 percent.

The Moon does not alter its distance from the Sun by much, but there is enough change to affect the intensity of light it receives. Also, the changing distance between the Moon and Earth affects the amount of sunlight reflected from the Moon to Earth. However, these changing values are rarely noticeable because of the interfering effects of the Earth's atmosphere, which diffuses the moonlight.

The greatest amount of moonlight comes during a full moon. At the highest point in the sky, a full moon gives about 0.02 foot-candles of

PHASE BRIGHTNESS

PHASE	AGE IN DAYS	PHASE ANGLE	RELATIVE BRIGHTNESS
New Moon	0	180°	0.00
	3	−143°	0.01
First Quarter	7.4	−90°	0.08
	10	−58°	0.22
Full	14.8	0°	1.00
	18	39°	0.37
Last Quarter	22.1	90°	0.08
	25	125°	0.02

Information adapted from various sources, including *The Strolling Astronomer: Journal of the Association of Lunar and Planetary Observers*. Brightness as indicated is based on average distances of the Sun, Moon, and Earth.

The first quarter moon and the last quarter moon are only about 8 or 9 percent as bright as the full moon.

BRIGHTNESS RULES
OF THUMB

The variation in brightness of full moons, not including effects of atmosphere or light pollution, is about 10 percent. The apparent size and brightness of the Moon is determined by a combination of factors, including:

◆ **The closeness of the Moon to the horizon.** Especially during full moons, the Moon appears much larger when it is just rising or setting because the proximity of the horizon acts as a visual guide to its size. When the Moon is high in the sky, there are nothing but stars and planets for comparison, both of which are just points of light next to it.

◆ **The angle of the Moon's path compared to the horizon.** This angle is greatest after sunset in the spring months and before sunrise in fall months. The angle is smallest before sunrise in the spring months and after sunset in fall months. When the angle is smallest, the Moon stays closer to the horizon for a longer period of time, adding to the impression of increased size.

◆ **The distance of the Moon from the Earth.** At its greatest distance (at the point of apogee), the Moon appears smaller, at its closest distance (at the point of perigee), it appears larger.

◆ **The speed of the Moon in its orbit around Earth.** At greatest speed (around the time of perigee) the phases appear to change more quickly; at slowest speed (around the time of apogee), the phases appear to change more slowly.

illumination. Quarter moons provide only about 8 to 9 percent of the light of a full moon.

These differences come from the spherical shape of the Moon and the coarse, uneven surface. Sunlight striking the full face of the Moon, as seen during a full moon, produces a maximum reflection because it is being reflected directly back toward the Earth. At any other time during the phase cycle, sunlight is striking the surface obliquely, limiting its ability to be reflected directly back at the Earth.

Recently, another unusual factor has been discovered that plays a key role in the production of moonlight. Examination of moon dust brought back from the Apollo expeditions has revealed the role of tiny particles that cling to the surface of lunar sand. In reflecting sunlight, these particles act to amplify rays of light, a condition labeled "coherent backscattering." Under certain conditions, such as during a full moon, the reflection intensifies, producing more visible light than during other phases.

LUNAR ECLIPSES

Lunar eclipses are produced when the Earth's shadow falls on the Moon. This would happen every full moon if the Moon orbited around the Earth in the same plane as the Earth orbits around the Sun. The Moon's orbit, however, is tilted about 5 degrees above the Earth-Sun plane. This tilt itself, however, rotates, allowing eclipses to happen when the tilt of this plane lines up with the Earth-Sun plane, blocking sunlight.

A lunar eclipse is visible over an entire hemisphere and is seen at the same time to everyone who is in sight of the full moon. Because of local time zones, however, the times of a lunar eclipse can span many hours. Lunar eclipses can last for more than three hours because the Moon and the Earth are moving slowly in relation to each other, and the shadow cast by the Earth is so large. Because of their sizes and the relative distances between the Earth, Moon, and Sun, this shadow is much larger than that cast by the Moon on the Earth (during a solar eclipse).

Although eclipses are always caused by the same general lineup of Sun, Moon, and Earth, each lunar eclipse may have its own unique visual characteristic. Colors and the deepness of the shadow on the surface are affected by the type of eclipse, local weather conditions, atmospheric conditions, and the geographic location of the observer. Volcanic activity may also affect the visibility of an eclipsing Moon, usually making it darker and less visible; sometimes the volcanic dust in the upper atmosphere may create unusual colors. During a total

The shadow cast during an eclipse has two components, a darker central area (the umbra) and a lighter outer area (the penumbra).

UMBRA   PENUMBRA

31

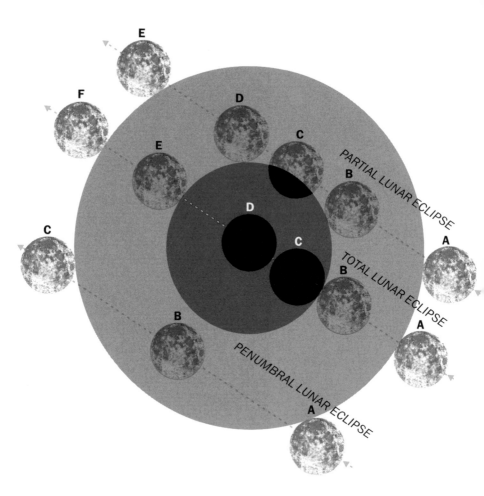

LUNAR ECLIPSE SEQUENCE

PARTIAL ECLIPSE
A Moon enters penumbra
B Moon enters umbra
C Middle of eclipse
D Moon leaves umbra
E Moon leaves penumbra

TOTAL ECLIPSE
A Moon enters penumbra
B Moon enters umbra
C Total eclipse begins
D Middle of eclipse
E Moon leaves umbra
F Moon leaves penumbra

PENUMBRAL ECLIPSE
A Moon enters penumbra
B Middle of eclipse
C Moon leaves penumbra

eclipse, the appearance can also be affected by solar conditions, particularly sunspot activity and the relative distance between the Moon and the Sun. A relationship is known to exist, for example, between the 11-year cycle of solar activity and the brightness of lunar eclipses, with the eclipsed moons dimmer when solar activity is low. For total lunar eclipses, some astronomers use a standardized scale, known as a Danjon lunar eclipse scale, to accurately rate the Moon's appearance during these events.

There are three kinds of eclipses of the Moon.

PENUMBRAL ECLIPSE

This is a partial eclipse, with the Moon in only the secondary shadow (penumbra) of the Earth. A penumbral eclipse is sometimes called an appulse eclipse. During a penumbral eclipse, the Moon's light is dimmed, but it does not go dark, because the penumbral shadow is not dark enough to block out all of the Sun's illumination. Often, there is no visible line separating the shadow from the sunlight on the Moon's surface, and the eclipse is only noticeable as a slight darkening of the lunar surface.

DANJON LUNAR ECLIPSE SCALE

0 Very dark. At the middle of totality, darkness almost completely obscures the Moon.

1 Color gray or brownish. Moon darkly shadowed with some features faintly visible.

2 Color rusty brown or dark red. Umbra is very dark in the center and lighter at the edges.

3 Color reddish to brick red. Edges of the umbra are lighter in color and may appear yellowish.

4 Color orange or copper. Edges of the umbra are very light and may appear bluish in color.

PARTIAL ECLIPSE

The Moon partially enters the main shadow (umbra) of the Earth.

ECLIPSE RULES OF THUMB

◆ Full moons are the only time lunar eclipses occur.

◆ New moons are the only time solar eclipses occur.

◆ A solar eclipse always occurs two weeks after or two weeks before a total lunar eclipse.

◆ Lunar eclipses can last for a maximum of 3 hours and 40 minutes, with the period of totality lasting for as long as 1 hour and 40 minutes.

◆ Solar eclipses can last for a maximum of 7 minutes and 40 seconds if they are total (at the equator), 12 minutes and 24 seconds at most if they are annular.

◆ Lunar eclipses can never happen more than three times a year. Solar eclipses happen at least twice a year but never more than five times a year.

◆ Lunar eclipses are visible over an entire hemisphere. Solar eclipses are visible in a narrow path that is a maximum of 167 miles wide (269 km).

◆ The greatest number of solar and lunar eclipses that can happen in a year is seven.

◆ At any specific geographic location on the globe, a total solar eclipse can occur only once every 360 years, on average.

◆ Solar eclipses and lunar eclipses go together in pairs. A solar eclipse is always followed or preceded by a lunar eclipse, within an interval of 14 days. Eclipses may also occur in threes, alternating lunar, solar, lunar.

◆ The characteristics of one eclipse are repeated every 18 years, 11 days, and 8 hours, with some minor variations. This long-term rhythm is called the Saros cycle. At any given time, there may be several dozen different series of this cycle in effect.

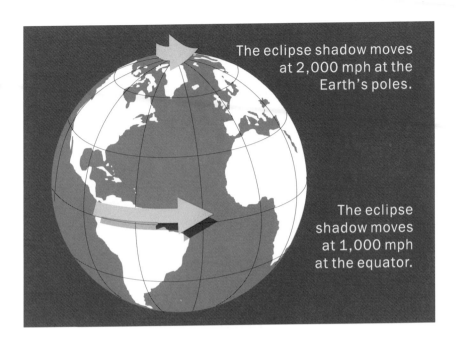

The eclipse shadow moves at 2,000 mph at the Earth's poles.

The eclipse shadow moves at 1,000 mph at the equator.

TOTAL ECLIPSE

The Moon is completely inside the main shadow (umbra) of the Earth. The dark umbral shadow cast by the Earth does not completely obscure the Moon but changes its color to a dull copper tone, an effect created by earthshine (light reflected off the Earth onto the Moon). The color is created by the filtering effect of the Earth's atmosphere, which removes all but the red wavelengths of sunlight. The Moon can stay in the umbral shadow of the Earth for as long as 90 minutes. The movement through the penumbral shadow can last for about 60 minutes.

SOLAR ECLIPSES

The Moon also causes eclipses of the Sun. When the Moon comes directly between the Earth and Sun — this can only happen during the new moon phase — it blocks out the Sun's rays. Depending on how the Moon and the Sun line up, there can be either partial or total blockage of the Sun's disk. Most months, the Moon's path is too high or too low as it crosses the Sun's position, so there is no eclipse activity.

A total solar eclipse occurs when the Moon completely blocks out the Sun. However, the elliptical orbit of the Moon can place it at varying distances from the Earth. When the orbit is closer to the Earth during a solar eclipse, the Moon appears larger and therefore blocks out the Sun for a longer period of time. When the orbit is farther from the Earth during an eclipse, the Moon appears smaller and may not quite cover the Sun's disk, allowing a thin ring of light around the eclipse shadow. This event is referred to as an annular eclipse.

A solar eclipse is much shorter in duration than a lunar eclipse because the Moon's shadow is falling on a rapidly rotating Earth. The maximum time for a solar eclipse is 7 minutes and 40 seconds (if it is located on the equator) but most of these eclipses are much shorter. The moving shadow cast by the eclipse is called an eclipse track and is usually about 3,000 miles long and 100 miles wide. The width can vary from almost nothing to a maximum of 167 miles wide.

The Moon's shadow during a solar eclipse moves along at speeds between 1,000 and 2,000 miles per hour. The slowest movement is at the equator and the fastest at the poles. The speed of the shadow is caused by the combined movements of the Moon and the Earth; the Moon is moving at about 2,000 miles per hour eastward and the Earth (at the equator) rotating at about 1,000 miles per hour.

There are three kinds of eclipses of the Sun.

PARTIAL ECLIPSE

The Moon is in conjunction with the Sun but its path does not take it directly across the center of the Sun's disc.

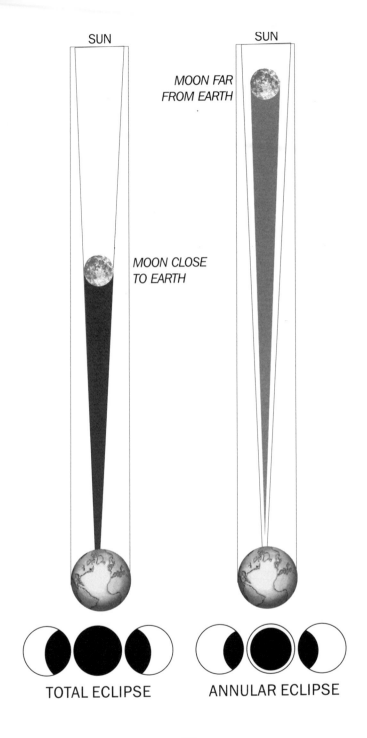

SUN

MOON FAR
FROM EARTH

MOON CLOSE
TO EARTH

TOTAL ECLIPSE

ANNULAR ECLIPSE

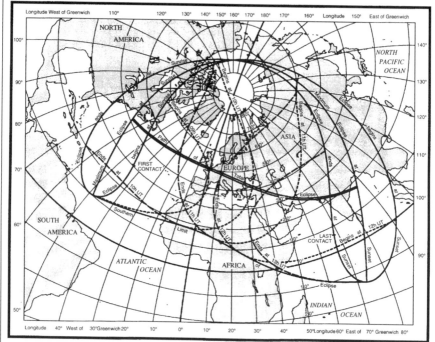

Eclipse maps plot the path of solar eclipses and are produced a few years in advance by the United States Naval Observatory and published by the U.S. Government Printing Office.

(Map reprinted from: *Astronomical Phenomena for the Year 1999*, published by the U.S. Naval Observatory.)

ANNULAR ECLIPSE

The Moon is in conjunction with the Sun, but is at a far point in its orbit around the Earth. The Moon's image is thus too small to completely cover the Sun's disc, and a ring of sunlight is visible around the edges.

TOTAL ECLIPSE

The Moon is in conjunction with the Sun and is at a close point in its orbit around the Earth. The Moon's image is thus large enough to completely block out the Sun's disc.

EARTHSHINE

The light of the Sun reflects not only off the face of the Moon, it also reflects off the Earth's surface. The reflected sunlight can produce visible light on the Moon, referred to as earthshine. During periods when the Moon is lit fully by sunlight — during the full moon — earthshine is not visible. During most of the lunar month and during eclipses the earthshine effect may be seen if atmospheric conditions permit. It is visible as a dark, grayish, dull copper, or reddish hue. During total lunar eclipses, this phenomenon may also be visible.

Just after the new moon, the emerging first crescent moon sometimes produces an effect referred to as the "old moon in the new moon's arms." This phenomenon is visible just after sunset a few days after the new moon when the young crescent moon is beginning to set in the west. A thin crescent-shaped illumination on the lower right side of the Moon's face will be seen, but the rest of the surface may also stand out from the surrounding sky with a copper or reddish hue.

For early risers, the same effect is sometimes visible as the old crescent moon sets just before sunrise. The best time of the year to experience this phenomenon is in late winter and early spring.

When the first crescent moon is only a few days old, light reflected back from Earth will sometimes illuminate the whole surface.

MOONLIGHT EFFECTS

A strong source of light such as the Sun or Moon can produce interesting optical effects when combined with the right atmospheric conditions. A rainbow — produced when sunlight is refracted through water droplets in the air — is perhaps the most familiar of these effects. When the light from the Moon is refracted through water droplets, a similar effect can be produced. This prismatic display is known as a moonbow or lunar rainbow, but is much less intense than a solar rainbow. Observers may see only the palest of colors. Moonbows are most likely to be seen when the Moon is full or within a few days of full.

When moisture is present high in the atmosphere in cirrus clouds, it is frozen into ice crystals. If the Moon is in the right position, the moonlight can form a halo or ring around the Moon. Ice crystals have six sides and normally refract light at an angle of 22 degrees, typically

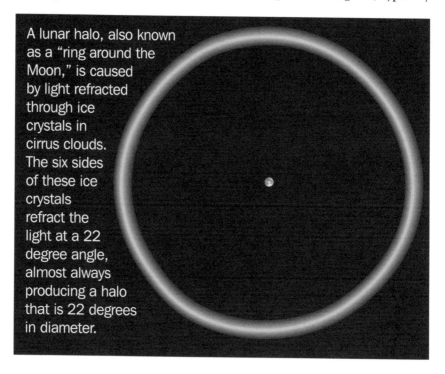

A lunar halo, also known as a "ring around the Moon," is caused by light refracted through ice crystals in cirrus clouds. The six sides of these ice crystals refract the light at a 22 degree angle, almost always producing a halo that is 22 degrees in diameter.

creating a lunar halo that is 22 degrees in diameter. Less typically, the halo may be produced by different angles in the crystals and appear 46 degrees in diameter. The same action that produces this ring also breaks the light into colors, just like a prism. The inner edge of the halo is blue and the outer is red when it is 22 degrees; the colors are reversed when it is 46 degrees.

Moondogs, also known as mock moons, are also produced by the interaction of moonlight and moisture in the atmosphere. With the right combination of humidity and angle, an observer may see a paler halo-like image of the Moon off to the side of the Moon itself. The official name for this effect is a paraselene if the extra image is seen 22 or 46 degrees away. If the image is at 90, 120, or 140 degrees, it is called a parantiselene. If it is at 180 degrees, it's an antiselene.

Moon pillars are another form of halo related to the Moon, but are rarely visible. Moon pillars can be seen when the Moon is near to the horizon, either just before or just after setting or rising. These are pale shafts of light above or below the Moon and are caused by moonlight reflecting off of ice particles or snowflakes in the atmosphere.

Another type of light show created by the Moon is the corona. Like a lunar halo, a corona is produced by high, thin clouds, but it is not as large in diameter. A typical lunar corona is only one or two degrees in diameter and closely fringes the Moon itself. A corona also may appear in several colors, like a solar rainbow, although the colors are not as intense. Typically, a slight reddish or bluish tint is visible. On rare occasions, two or more coronas in a concentric pattern may be produced at the same time. A lunar corona can also appear along with a lunar halo.

THE MOON IS FLAT?

Early observers of the Moon, including Galileo, noticed that the full moon appeared to be flat. If a round object such as a balloon or ball is illuminated like the full moon, a distinct three-dimensional effect occurs. In fact, almost any sphere exhibits this common effect from illumination — a gradual darkening around the edges with the brightest area in the center. The Moon does not.

That the full moon has almost equal illumination everywhere across its surface is the result of its unusual surface texture. Light from the Sun striking the Moon is almost completely absorbed by the surface. Only about 7 percent of the sunlight is reflected. The percent of light reflected from an object is referred to as albedo, designated as measurement from 0 to 1, with 1 being 100 percent reflection. The Moon's albedo is 0.07. In comparison, the albedo of Earth is about 0.30 (30 percent); for Venus it is 0.56; and for Neptune it is 0.73.

The texture and color of different substances affects the albedo. For instance, on Earth the albedo of concrete ranges from 0.17 to 0.27; the albedo of snow is 0.45 to 0.90; the albedo of deserts is 0.25 to 0.30; and the albedo of soil is 0.05 to 0.15. Various features on the Moon also have different albedos. This results from the primary materials present at each location. The range is from 0.05 (Sinus Medii) to 0.176 (Aristarchus). Even during eclipses, the lunar features with the highest albedos can be seen because of earthshine.

The almost complete absorption of light on the Moon is caused by the rough, uneven texture of the surface, and a layer of material made of particles with measurements less than a half inch in diameter. The particles of this lunar soil are themselves coated with a fine layer of rock powder that efficiently scatters the incoming sunlight. Around the edges of the illuminated disk of the full moon, this redirection of light almost eliminates the shadows that would ordinarily be seen on the sides of a typical spherical object, producing the visual effect of roundness. When the Moon is not full, the light reflected from its surface is not aimed directly at Earth, diminishing the scattering effect. Thus, anytime but the full moon is a good time for observing surface features.

FIRST SIGHTING

Every lunar month begins with the phase of the new moon. This phase, however, provides little interest for observers unless there is an eclipse. Otherwise, by definition, the Moon is too close to the Sun for viewing. Within one or two days, however, the first crescent moon appears in the western sky just after the Sun sets, providing a striking visual event.

For some sky watchers and amateur astronomers, much effort and time is spent attempting to sight the first crescent moon at the earliest possible time after the new moon. This task is not as easy as it seems, as visibility of the slender, illuminated crescent is affected by a number of variable factors, including local humidity (which produces haze in the atmosphere), weather conditions, the time of year, the geographical location of the viewer, and the time of day when the new moon occurs.

The current record for the earliest crescent moon spotted with the naked eye is 15 hours and 30 minutes. Using binoculars, a first crescent moon has been seen at 13 hours, 32 minutes. With a telescope, the earliest sighting is only 12 hours, 7 minutes after the new moon.

Ordinarily, the first crescent moon will be easily visible within three calendar days of the new moon, and in most months, the first crescent can be sighted by the second calendar day after the occurrence of this phase. But in order to spot the emerging crescent sooner, extra care and planning is required.

The major problem associated with seeing the crescent moon is its lack of illumination. Because so little of the surface of the Moon is lit and the illuminated crescent is backlit by a sky that is not yet dark, it provides less of a visual target for the human eye. This illumination is produced by the nearness of the Moon to the Sun; it is just leaving the point in time of the new moon, when it is as close to the Sun as it gets every month. This means that during this period — from a few days before the new moon to a few days after the new moon — the Moon's position in the sky is close to the Sun and immersed in the light that it produces.

The actual illumination of each crescent moon varies from month to month, a function of the angle between the Moon and the Sun, called the elongation or arc of light. When this angle is zero degrees, the Moon may pass in front of the Sun, producing an eclipse, but most months do not produce eclipses because the angle is greater than zero.

In non-eclipse months, elongation varies from one to five degrees.

Another variable that affects the amount of illumination is the apparent size of the Moon, which varies according to how far away from Earth it lies. This distance changes over time because of the elliptical shape of the Moon's orbit; when the Moon is closest to the Earth — during perigee — it is largest, providing a bigger target for observers.

The same conditions that produce a first crescent moon are duplicated for an old last crescent moon. During this phase, however, the Moon is gradually being overtaken by the Sun and creeps closer and closer to it from day to day. Instead of being an evening phenomenon, it is visible in the morning just before the Sun rises. And instead of being up and to the left of the Sun, as seen from Earth, it is up and to the right.

Based on astronomical calculations and simulated perfect conditions, the earliest that a crescent moon could be seen from the Earth is when there is more than 7 degrees of elongation. Elongation is the angle between the Sun and the Moon, and is related to the age of the Moon and the tilted plane of the Moon's orbit relative to the Sun. At an extreme, the Moon may be as much as 5½ degrees from the Sun at the instant of the new moon; in order for a first crescent to be visible, it must be at least a few degrees more. Even at the instant of new moon, without the Earth's atmosphere spreading and deflecting the sunlight, it is theoretically possible to see the shape of the Moon because of the light reflected onto its surface from the Earth.

In fact, this was accomplished in 1966 when a photograph was taken

---

RECORD SIGHTINGS

The earliest that first crescent moons have been sighted:

15 hours, 30 minutes (naked eye)

13 hours, 32 minutes (binoculars)

12 hours, 7 minutes (telescope)

from a special camera mounted in a rocket launched from the White Sands Missile Range in New Mexico. The photograph, taken when the Moon was only 2 degrees from the Sun, showed a small, irregular illumination on the side of the Moon facing the Sun.

For Earth-bound viewers, however, the quest to sight a first crescent moon has existed long before space flight, rockets, or even telescopes. Many ancient calendars, in fact, were based on the movements of the Moon, and human observation was required to keep track of the beginning of lunar months.

In the modern world, a few calendars are still based on lunar rhythms. Fundamentalist Muslims, for example, follow the traditional Islamic calendar, and must sight the first crescent moon in order to mark the beginning of calendar months. This sighting must be done without the aid of magnification and two reliable witnesses to each sighting are required in order for the new month to begin. Even with the aid of sophisticated computer programs that can closely predict when this event might happen, a physical sighting is still needed.

At its earliest stages, the first crescent moon is a striking sight. If the Moon were smoothly spherical, the lit area would extend close to one-half the distance around the circumference, or 180 degrees. But because the surface is rough and irregular, the arc that is illuminated

For each additional calendar day after the new moon, the young crescent moon appears another 13 degrees further up in the sky.

When the new crescent moon is less than 24 hours old, it is still close to the Sun and is extremely difficult to see with the naked eye because the sky is still too light for contrast.

FIRST CRESCENT
SIGHTING GUIDELINES

◆ Plan ahead by looking up times for new moons in almanacs, local newspapers, or the Moon Calendar (and translate these times to your local time zone). To spot the earliest first crescent, choose those dates when these times are 12 to 23 hours in advance of sunset at your location.

◆ Optimum time target: 10 to 60 minutes after sunset.

◆ Best sighting is from late winter through early spring, when the ecliptic (apparent path of the Sun and Moon across the sky) is steepest relative to the horizon. Thus, the Moon can be closer to the Sun just when the sky gets dark enough to spot the emerging crescent.

◆ Best sighting conditions are seasons of the year when humidity is low. Usually, that means winter months are better than summer months. Smog and overcast weather conditions are also to be avoided.

◆ Best locations are where humidity is low. Usually, that means southwestern states in the United States are better than midwestern, southern, and eastern states.

◆ Best locations are those with the greatest expanse of unobstructed western horizon.

◆ People with better eyesight will have better luck than people with diminished eyesight.

◆ Locations with higher altitudes provide better conditions than locations with lower altitudes.

◆ Locations with low latitudes provide better conditions than locations with high latitudes.

◆ Best sighting opportunities are within a few days of lunar perigee (Moon nearest to Earth).

◆ The chance of sightings may be improved by first using binoculars to locate the emerging crescent. Warning: do not attempt to look in the direction of the visible Sun with binoculars.

will be somewhat shorter, typically about 130 degrees but as short as 30 degrees when the crescent is extremely new. And this narrow sliver of surface is not an even, symmetrical "slice," but dotted and broken along its inner edge from the effects of shadows of craters, valleys, and mountains.

Close to the ends of the arc, the cusps are also not perfect, but ragged and broken, further irregularity caused by the lunar landscape. Here, high elevations, such as mountain tops, catch the sunlight and form bright dots and small splotches amid darker areas.

When very young, the new crescent moon is silhouetted against a sky that is not yet dark. This diminishes another notable lunar effect. Earthshine, sunlight reflected off the Earth's surface and back to the Moon, ordinarily strikes the complete visible surface of the Moon, even that part not illuminated directly by the Sun. The effect produces the distinct, although dim, outline of the lunar sphere. Near the horizon and viewed against a bright sky, however, earthshine is less likely to be visible and may not appear at all if the crescent moon is less than 24 hours old.

OBSERVING
THE FULL MOON

The full moon occurs at a precise moment in time. In astronomical terms, it is full at the moment when it is directly opposite the Sun. If that moment were exactly at the time of sunset for your location, you would see the full moon rise in the east just as the Sun was setting in the west. This rarely happens, however. In almost every month, what usually occurs is a discrepancy in the time at which the Moon is full and the time it rises at your location.

The consequence is that the fully lit Moon will rise from a few minutes to an hour before or after the local time of sunset. The Moon, however, will appear full for at least a day before and after the exact moment when it is astronomically "full" — and sometimes for two or three days — because its spherical shape does not immediately begin to show the shadow lingering on the left side just before full moon or the shadow on the right side just after full moon. The period of apparent fullness is related to how fast the Moon is traveling in its orbit and the angle of the Moon's orbit compared to the Sun. Both of these vary from month to month.

On the marked day of a full moon, no matter what the astronomical time of the event, the Moon will appear full during the evening hours closest to the astronomical time of the phase. In general, if the marked time is 9:00 A.M. or later, you will see the fullest Moon on the evening of that day; if the marked time is earlier than 9:00 A.M., you would see the fullest Moon on the previous evening.

No matter where you are on the globe, the time listed as Universal time (UT) — also known as Greenwich Mean Time (GMT) — for any phase of the Moon is the exact time when that phase occurs. The time of full moons and new moons are usually listed in calendars and almanacs in this time format. This is the official local time at the Greenwich Meridian (0 degrees longitude, near London, England) that is used as a worldwide standard by astronomers.

There are no variations in latitude or longitude that affect the time when phases occur, but if you are not in the UT zone, you do have to convert that time to your local time zone. In order to convert to local time, you must add or subtract the right number of hours, determined by how many time zones are between you and Greenwich, England. (See chart on page 123.)

Another factor that confuses the date when moon phases occur comes from local reporting. Most, but not all, newspapers and radio and television stations use phase times that have already been converted for their region. But some almanacs, international publications, and other media may use Universal Time when mentioning a particular moon phase, eclipse, or "blue moon" (for more on blue moons, see page 112). Although errors may crop up in reporting, most of the time mistakes made about moon phases come from this time zone confusion.

Traditionally, full moons in fall months have been considered the biggest, brightest of the year. The harvest moon, for example, was so named because it provided additional hours of light during which farmers were able to complete their annual harvest. The harvest moon, not coincidentally, is linked to the month of September, a month when many crops have traditionally been harvested, and also when a major astronomical event occurs every year in the northern hemisphere— the fall equinox, also referred to as the autumnal equinox. This event marks the point in the Earth's orbit around the Sun when the ecliptic — the apparent path of the Sun's movement across the sky — is halfway between its annual movement north and south. On this day, September 23 or the day before, the length of day and night are almost equal, hence the name. Also at this time, the full moon rises at its shallowest angle to the horizon, making it loom larger and amplifying its image.

At other times of the year, the Moon rises about one hour later from one day to the next. But close to the autumnal equinox, the angle of the Moon's path cuts this daily lag down considerably. From August through October — peaking in the crucial month of September — the Moon may rise only about 15 minutes later each day, compounding its already powerful image. When the full moon occurs very close to the equinox, the effect is magnified even further.

LUNAR TILT

The Moon appears to be tilting, or changing position, as it crosses the sky. This visual effect is most apparent when the Moon is waxing, and the lighted portion of the surface appears to turn from pointing to the west (right) to almost straight down as the Moon sets. The Moon is not actually turning, however, but is following a curved path across the sky, the path defining its orbit. Pictures or illustrations of the lunar phases usually show the Moon when it is at its zenith (highest point above an observer) and therefore oriented straight "up and down" relative to the earth. The illustration below shows the actual visual appearance of the Moon as it rises and sets on the same night. Variations in this tilt are affected by the latitude of an observer. The farther north, the greater the affect. Below the equator, the Moon is still affected by this characteristic of its orbit, but because observers are seeing it "upside down" compared to the northern hemisphere, its image is also reversed from top to bottom.

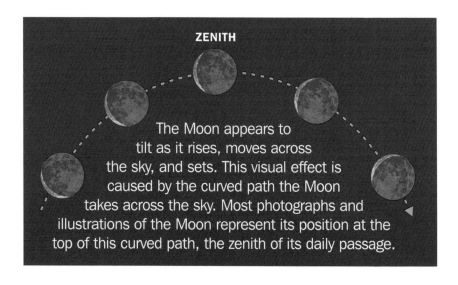

ZENITH

The Moon appears to tilt as it rises, moves across the sky, and sets. This visual effect is caused by the curved path the Moon takes across the sky. Most photographs and illustrations of the Moon represent its position at the top of this curved path, the zenith of its daily passage.

OCCULTATIONS

The sky is full of objects visible to observers on Earth. Planets and stars are visible as points of light to the naked eye. The planets appear to move at different rates than the stars as they are much closer to the Earth, and stars are not actually moving at all; their apparent movement is due to the movement of the Earth. Both stars and planets, however, are not moving at the same rate as the Moon and a star or a planet will occasionally be obscured by the Moon when their paths overlap in the sky. This phenomenon is called an occultation.

Occultations occur whenever one body crosses in front of another body. Occultations caused by the Moon are the most obvious as the Moon's image is the largest in the sky. The most observable occultations to watch are those that occur when the Moon is less than full because there is less light to interfere with the event.

Observers on Earth see an occultation as a point of light approaching the Moon from the left. The stars and planets move across the sky faster than the Moon, accounting for this apparent approach. Occultations can last for a few minutes or up to an hour, depending on the Moon's path (declination), that is, whether the object appears to intersect the Moon at the widest part or just "grazes" the edge. At the instant of occultation, the point of light will suddenly disappear, appearing again when it emerges on the other side of the Moon. Grazing occultations appear as a blinking or twinkling phenomena, with the mountains and ridges on the Moon's surface momentarily breaking up the light from the star. When the large planets overtake the Moon, they may take a few seconds or longer to be occulted because of their size.

The brightest stars are also the most visible under the widest range of viewing conditions, but only those stars that lie in the zone of the Moon's path — about 5 degrees on either side of the ecliptic — can be intercepted by the Moon. Major stars likely to be observed being occulted by the Moon are Aldebaran, Antares, Regulus, the Pleiades, and Spica.

Lunar occultations are predictable because of the known orbital characteristics of the Moon. However, the times of occultations are difficult

During a grazing occultation, a star will appear to "blink" on and off as it moves behinds peaks and valleys along the edge of the Moon.

to compute because of the varying path of the lunar orbit and the range of possible locations of observers. Only the position of the stars remains unchanged. A distance of a few hundred yards to a mile north or south on the Earth's surface can make the difference between seeing a grazing occultation or a "near miss." The observer's longitude — east or west — determines the time when an occultation may be observed; the further to the east, the later the event.

Amateur astronomers who are serious about observing lunar occultations can receive specific information about their locations from the U.S. Naval Observatory. In return, they are expected to keep notes on their observations and share this information with the Observatory. Serious inquiries about this program can be addressed to U.S. Naval Observatory, Washington, D.C. 20390.

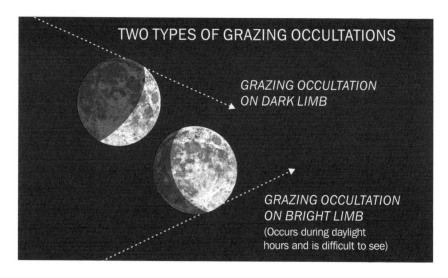

TWO TYPES OF GRAZING OCCULTATIONS

GRAZING OCCULTATION ON DARK LIMB

GRAZING OCCULTATION ON BRIGHT LIMB
(Occurs during daylight hours and is difficult to see)

LIBRATIONS

The Moon always shows the same face to the Earth. Earth observers, however, get to see more than half of the surface of the Moon because of phenomena called libration. At any one time, the most that can be seen of the surface is only 41 percent because the spherical shape of the Moon hides the area close to the perimeter. But librations allow the front surface of the Moon to be seen from slightly different angles at different times, producing an overall picture of the lunar surface that adds up, over time, to 59 percent of the total.

Librations are measured using longitudinal and latitudinal coordinates. Both are figured from a central point that is at a fixed geographical location on the lunar surface. This point is in the Sinus Medii, a small flat plain just below and to the right of Copernicus, a large rayed crater that is visible to the naked eye. Two meridians, one north and south (the Central Meridian) and one east and west (the Lunar Equator), cross at this point, which is also used to mark locations on maps of the lunar surface.

Different librations affect different sides of the Moon, with each contributing different degrees of added surface area. Depending on the orbital characteristics, the effect may vary from day to day and month to month. Sometimes, librations also overlap, creating an even stronger effect. The maximum added area that can be seen from a combination of librations is a little more than 10 degrees.

LIBRATION IN LONGITUDE

This libration is produced by the elliptical orbit of the Moon. If the orbit were a circle, the Moon's face would always point directly at the Earth and its orbital speed would remain the same. However, due to the nature of an elliptical orbit, the speed of the Moon changes depending on which part of the orbit it is in. This is also true of any object moving in an elliptical orbit. When moving from its fastest point (closest to Earth) to its slowest point (farthest from Earth), the Moon's speed is slowing down, but its rotation speed remains the same. For a period of time, the face of the Moon is therefore not pointed directly at the Earth. This "lag" effect allows observers to see an extra slice of surface,

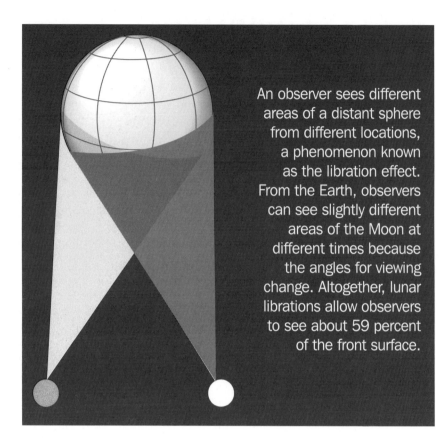

An observer sees different areas of a distant sphere from different locations, a phenomenon known as the libration effect. From the Earth, observers can see slightly different areas of the Moon at different times because the angles for viewing change. Altogether, lunar librations allow observers to see about 59 percent of the front surface.

in effect, to "peek" around the edge of the Moon. When the Moon is at the 90 degree point in its revolution (one-fourth of the way around the Earth), it is 97 degrees through its rotation. This libration is called longitudinal because the extra surface areas are exposed along lines of longitude (perpendicular to the equator). The total extra surface that can be seen with this libration is 8 degrees (7° 57', to be exact).

LIBRATION IN LATITUDE

The plane of the Moon's orbit is tilted about 5 degrees away from the plane of the Earth's orbit around the Sun. For half of a lunar cycle, the Moon is below the ecliptic, and for the other half the cycle it is above the ecliptic. Each of these half cycles exposes an extra "slice" of the

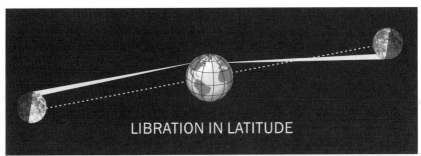

LIBRATION IN LATITUDE

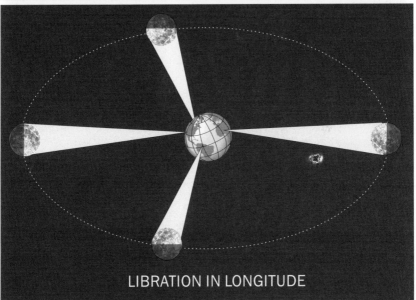

LIBRATION IN LONGITUDE

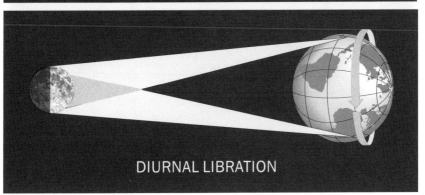

DIURNAL LIBRATION

lunar surface at the top of its northern hemisphere or the bottom of its southern hemisphere, in effect allowing a "peek" above or below the normal limit of visible surface. This libration is called latitudinal because the extra surface areas that are exposed are great circles that are parallel to the equator. The total extra surface that can be seen with this libration is about 7 degrees (6° 51').

DIURNAL LIBRATION

Observers can "see over the top" of the Moon when it is rising, and "under the bottom" when it is setting. This is possible because the radius of the Earth adds an extra 4,000 miles of height advantage for looking "over" or "under" the Moon when it is on the horizon. This libration is called diurnal because it occurs every day, but only accounts for an extra 1 degree of visible surface.

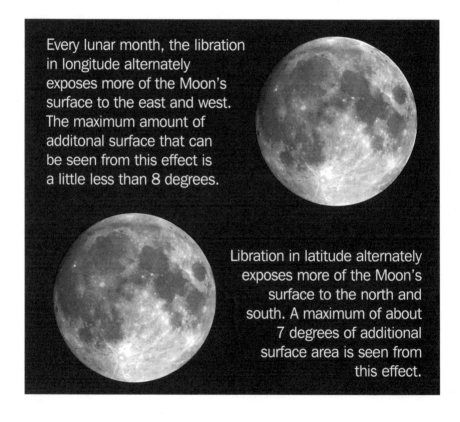

Every lunar month, the libration in longitude alternately exposes more of the Moon's surface to the east and west. The maximum amount of additonal surface that can be seen from this effect is a little less than 8 degrees.

Libration in latitude alternately exposes more of the Moon's surface to the north and south. A maximum of about 7 degrees of additional surface area is seen from this effect.

TIDES

The Moon is the nearest celestial neighbor to Earth and exerts a constant influence through gravitational attraction. This force is only one ten-millionth of the gravitational force of the Earth itself, but combined with other forces — including the centripetal force created by the spin of the Earth — it is one factor that produces tides on the world's bodies of waters. There are also tides created in the atmosphere and, to a much lesser degree, in the Earth itself.

Lunar gravity does not work alone in producing tides. They are also influenced by the centrifugal force created from the Earth's revolution around its barycenter (see page 1) and the gravitational attraction of the Sun. Gravitational attraction varies inversely with the square of the distance between two bodies. The Sun's mass, for example, is 27 million times larger than that of the Moon, but it is 390 times farther away from the Earth than the Moon. The result is that the Sun's gravitational force on the Earth is only 46 percent as much as the Moon's, making the Moon the most important factor for tides.

Tides, however, are not created by the direct pull of the Moon's gravity. The gravitational force of the Moon is tugging upward on the water, while the gravitational force of the Earth, which is far stronger is pulling down at the same time. Instead, water rises in tides because of a net balance of forces — the Earth pulling in and the Moon pulling out — averaging more in favor of the Moon. It does not happen in a perpendicular direction, however, but shows up where the external influence has a greater effect, from the side.

This type of gravitational attraction is known as tractive force. In a simplistic sense, it is the same phenomenon experienced if a person attempted to pick up a heavy object from above. In order to lift the object, it requires enough strength to completely overcome the weight, or mass, of the object. But if a person pulls the same object from the side, sliding it over a surface, for example, it takes much less force to move it.

Tractive forces create a "piling" effect on the Earth's oceans, pulling water toward the Moon from around the planet, adding up to the greatest effect — and the highest pile — closest to the spot that is underneath the Moon's position, its zenith. Because of the motion and relative postions of the Earth, the Sun, and that of the Moon, this pile, or high tide, is typically somewhat ahead or behind the actual zenith of the Moon. The time it takes for this high water to arrive before or after the Moon passes overhead is called the lunitidal interval.

At the same time that water has piled up on the Moon's side of the Earth, a second bulge has piled up on the opposite side of the planet. This second bulge is a result of water moving to create an equilibrium between the gravitational force of the Moon and the centrifugal force created by the movement of the Earth. Opposite from the Moon, the net effect on the oceans is to pull away from the Earth. Because the Moon's orbit around the Earth is completed about once every 25 hours, each of the two tidal peaks — as well as the two tidal troughs — occur halfway through that period, about 12½ hours apart.

The Sun's gravity also produces daily tides. But because the force is smaller than that of the Moon, the effect produces smaller tidal bulges. The Sun's daily cycles are also slightly shorter than the Moon's, with only 12 hours between solar tide peaks.

When the Sun, Moon, and Earth are in a line, the gravitational effect of the Sun adds to that of the Moon, creating maximum tides. This

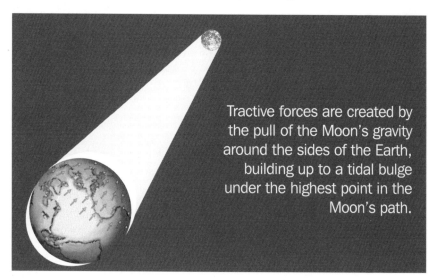

Tractive forces are created by the pull of the Moon's gravity around the sides of the Earth, building up to a tidal bulge under the highest point in the Moon's path.

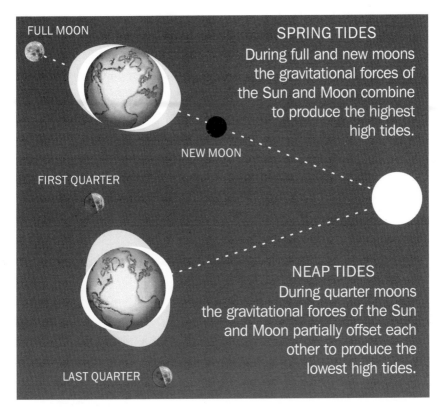

FULL MOON

SPRING TIDES
During full and new moons
the gravitational forces of
the Sun and Moon combine
to produce the highest
high tides.

NEW MOON

FIRST QUARTER

NEAP TIDES
During quarter moons
the gravitational forces of the Sun
and Moon partially offset each
other to produce the
lowest high tides.

LAST QUARTER

kind of alignment happens twice every lunar month because it is the same alignment that produces a new moon and a full moon. If the Sun and Moon are on the same side of the Earth (new moon) or on opposite sides (full moon), a spring tide is the result. Spring tides create the highest high waters and lowest low waters every month. When the Sun and Moon are at right angles (first and last quarter moons), a neap tide is the result, when the difference between high and low waters is at a minimum. When the Moon is closest to the Earth every month, gravitational forces are also greater, but the greatest effect on tides only comes when this orbital event coincides with a full or new moon, producing what is known as a perigean spring tide.

Tides are made even more complicated by the effects of local geography. The speed and height of tides are affected by water depth, wind, and the obstructions on shorelines and below water. In the open ocean, the total difference between high and low tides is about one foot. Along some coasts, the difference can be more than twenty feet.

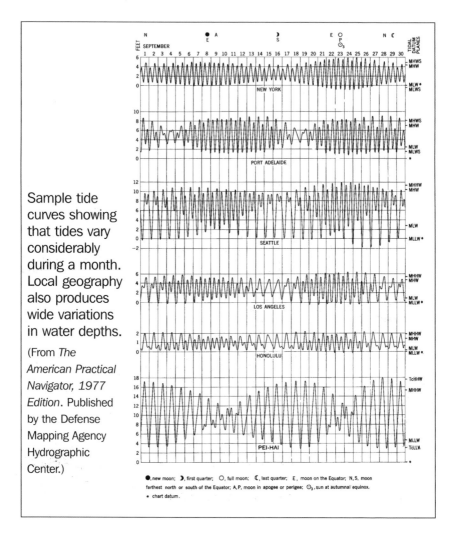

Sample tide curves showing that tides vary considerably during a month. Local geography also produces wide variations in water depths.

(From *The American Practical Navigator, 1977 Edition*. Published by the Defense Mapping Agency Hydrographic Center.)

PHOTOGRAPHING THE MOON

The Moon is the biggest and brightest object in the night sky, and attracts the interest of many photographers. But even when the Moon is near the horizon, and seems to be quite large, it is difficult to capture on film. The most common problem photographers have with this celestial object is its size. With a standard 50mm lens, the image on a negative or slide will be less than 1/50 inch in diameter (about .45 mm). Even if the image is enlarged 15 times, it will still only about ¼ inch in diameter (about 7 mm).

A longer lens will give better results, but use a tripod because of the longer exposure times needed. To determine the final size of the Moon, use this formula:

$$\text{image size on film} = \frac{\text{focal length}}{109}$$

For example:	LENS SIZE	IMAGE SIZE
	100mm	0.92mm
	200mm	1.83mm
	300mm	2.75mm

You can also use converters to improve the power of your lens. A x2 teleconverter will double the focal length of a telephoto lens. However, the f-ratio is also doubled, requiring longer exposures. For example, a 100mm lens at f/8 becomes 200mm at f/16 using a x2 teleconverter.

Exposure time for lunar photos is an important consideration, because the Earth is rotating while the shutter is open. Clock drives that synchronize telescopes with stellar motions can be used with some cameras to minimize this motion.

IMAGE SIZE	FOCAL LENGTH
	50mm
	135mm
	200mm
	300mm
	400mm
	500mm
	1000mm
	2000mm

Without a clock drive, the maximum exposure time for a fixed camera can be figured from this formula:

$$\text{exposure (in seconds)} = \frac{250}{F \text{ (focal length in mm)}}$$

The following formula allows you to calculate the exposure time needed for different types of film and different phases of the Moon. The film speed is indicated by A (ISO number), f is the f-stop number, and B represents a value corresponding to the brightness of the Moon (10 for thin crescent, 20 for wide crescent, 40 for quarter moon, 80 for gibbous moon, 200 for full moon).

$$\text{time (in seconds)} = \frac{f^2}{(A \times B)}$$

There are several methods for photographing the Moon using a telescope. With any of these methods, it is fairly simple compared to photographing other astronomical objects, and almost any kind of film will give satisfactory results. The main problem is the great contrast range in the image, from bright sunlight to deep shadow. The exception is the full moon, which is very bright with little contrast. The best film is one that has great exposure latitude — it can tolerate considerable over- or under-exposure.

ONLINE FILM ADVICE

Kodak http://www.kodak.com/
Fuji http://www.fujifilm.com/

Mistakes can then be compensated for in developing and printing, whether you do it yourself or have it done by a lab (black and white films are the best to use for this). It is best to choose one or two films and stick with them so you can learn their qualities. Some standard films are: Kodak Tri-X Pan (400 ASA), Plus-X (125 ASA), T-Max 100, and T-Max 400.

Color films have much less latitude than black and white. If you can make your own prints, use color negative films and manipulate them in the darkroom. Any of the Kodacolor VR or the Kodacolor VRG print films are good. Color reversal (slide) films have almost no exposure latitude, so bracketing is particularly important. Kodachrome 64, Ektachrome 64, and Ektachrome 400 are traditional favorites; Kodak

RULES OF THUMB FOR PHOTOGRAPHING THE MOON

◆ Always use a tripod.

◆ Photos of the full moon are flat and featureless. For more interesting pictures, photograph the Moon at crescent or quarter phases when the mountains and craters are illuminated from the side and cast shadows.

◆ Always bracket exposures since exposure times given by formulas are approximate, varying according to the exact phase of the Moon, atmospheric conditions, etc. To be safe, bracket at one and preferably two stops on both sides of the exposure suggested by the formula.

Ektachrome Elite II 50 and Fuji Velvia (ISO 50) are new choices. If you bracket carefully (two stops under and one stop over with slide film), you are almost certain to get a few excellent shots from a roll.

To shoot lunar eclipses, some special considerations apply. The biggest factor is a dramatic change in brightness, varying from +4 to –12. To compensate, film speeds between ISO 200 and 400 are recommended.

Three basic methods provide options for astrophotography.

PRIME FOCUS METHOD

This is the simplest method for lunar photography. If you have a catadioptic telescope, such as a Schmidt-Cassegrain or Maksutov, remove the camera lens, and using a simple adapter (T-ring, available at camera stores and astronomical supply houses), attach the camera directly to the telescope (without its eyepiece). The telescope becomes the camera lens, in effect, a very long telephoto lens. For example, a 5-inch Celestron or Meade Schmidt-Cassegrain telescope becomes a 1250mm, f/10 telephoto.

AFOCAL METHOD

This method requires two tripods, one for the camera, and one for the telescope. The camera (with lens) is set up as close to the telescope eyepiece as possible, and focused on the image seen through the camera. The exposure time is calculated by determining the focal length (F) and the f/ratio of the telescope system.

F = focal length of camera lens x magnification of telescope

f/ratio = F/diameter of the telescope objective

Magnification of the telescope = $\dfrac{\text{focal length of telescope}}{\text{focal length of eyepiece}}$

PROJECTION METHOD

This method requires the use of a special adapter which is much more expensive than the simple T-ring used in the prime focus method. The adapter connects the camera without its lens to the eyepiece of the telescope. This is an advanced method used in most amateur astropho-

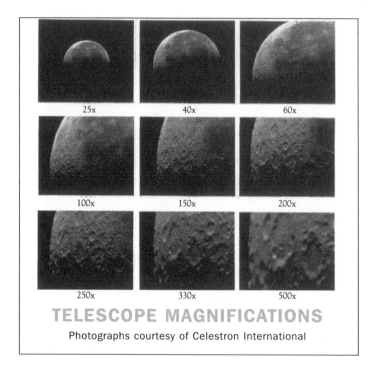

TELESCOPE MAGNIFICATIONS
Photographs courtesy of Celestron International

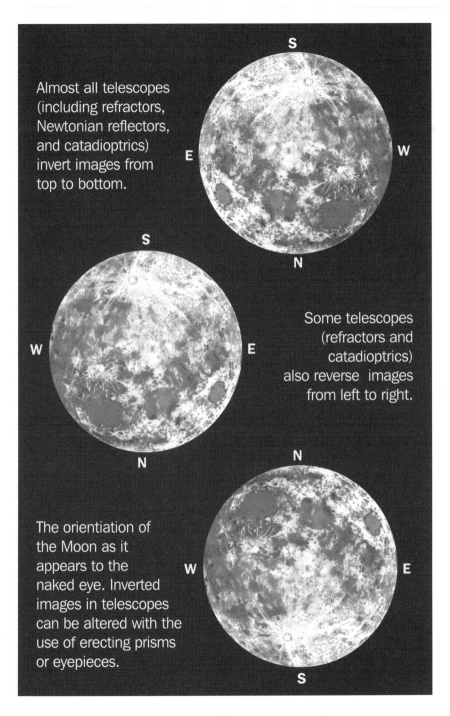

Almost all telescopes (including refractors, Newtonian reflectors, and catadioptrics) invert images from top to bottom.

Some telescopes (refractors and catadioptrics) also reverse images from left to right.

The orientiation of the Moon as it appears to the naked eye. Inverted images in telescopes can be altered with the use of erecting prisms or eyepieces.

DIGITAL MOONS

The image of the lunar surface shown on page 68 ("Major Visible Features") and several other Moon shots used in this book were created with the latest digital technology and a relic from the early days of the Apollo program. In the early 1960s, a 24-inch reflector was designed at Caltech as part of a NASA research project. This telescope was used to investigate the depth of lunar dust on the Moon's surface to avoid unpleasant surprises for the upcoming Apollo program. Now upgraded and back in use, this pioneering lunar instrument is coupled with a CCD camera and operated at the Mount Wilson observatory by the Telescopes in Education (T.I.E.) program. On March 12, 1998, Colleen Gino and Scott W. Teare, amateur astronomers working with the T.I.E. program, exposed a series of eighteen shots of the full moon. The shots were combined into a mosaic with graphics software (Microsoft Paint and Corel PhotoPaint), generating a master TIFF file.

The telescope: 24" f/3.5 Newtonian, masked with a 2" aperture

The camera: SBIG ST-6 CCD

The software: *TheSky Astronomy Software*

The shot: 18 images, 0.03 seconds exposure each

For information about the image: cgino@worldnet.att.net

For information about the Telescopes in Education program:
626-793-3100   http://tie.jpl.nasa.gov/tie/index.html/

For information about the *TheSky Astronomy Software*:
Software Bisque 303-278-4478   http://www.bisque.com

For information about the CCD camera:
Santa Barbara Instrument Group 805-969-1851
http://www.sbig.com

tography of planets, double stars, and deep-sky objects involving very long exposures and special films.

Beginning in the 1990s, digital cameras and image processing with personsal computers have begun to make an impact on the world of photography, for both amateurs and professionals. In astronomy applications, digital images are likely to gradually attract more fans as the price of equipment drops and processing power increases. Particularly with the use of the Internet to publish and exchange information, digital images will be the format of choice.

Photographs taken with traditional film methods can be turned into digital images by scanning. For the majority of applications in the graphics industry, the format of choice is TIFF (or TIF, for tagged image format file), but a variety of other formats also work well, and most can be translated easily to a different format. Each graphic format used to capture images translates the image's distinctive elements into individual bits, creating a bitmapped version of the original. Using a graphics program, these digital bits can be manipulated, adding contrast or brightness, removing unwanted elements, or otherwise altering the original. Such files can also be compressed, speeding up transmission times and reducing the amount of storage space required.

Although existing photographs can be turned into digital files, most amateur astronomers are more excited about capturing images from space directly in a digital format. Cameras that can record digital images are referred to as CCD instruments. CCD stands for charge-coupled device; a CCD camera projects an image onto a specialized computer chip (referred to as a CCD chip), which records variations in light intensity as a set of digital measurements.

CCD RESOURCES

The Art & Science of CCD Astronomy, by D. Ratledge, editor. 1997, Springer Verlag.

CCD Camera Cookbook, by Richard Berry. 1994, Willmann-Bell.

CCDSoft (software). 1997, Software Bisque

A Practical Guide to CCD Astronomy, by Patrick Martinez, Alain Klotz, and Andre Demers. 1998, Cambridge University Press.

N

PLATO   MARE FRIGORIS

MARE
IMBRIUM

POSIDONIUS

ARISTARCHUS

MARE
SERENITATIS

MARE
CRISIUM

COPERNICUS   MARE
VAPORUM

MARE
TRANQUILLITATIS

KEPLER

W

GUTENBURG

OCEANUS
PROCELLARUM

PTOLEMAEUS

MARE
FECUNDITATIS

GASSENDI

MARE
NECTARIS

MARE
NUBIUM

MARE
HUMORUM

TYCHO

S

MAJOR VISIBLE FEATURES

Photograph courtesy of Colleen Gino; for information on how this photograph was created, see "Digital Moons," page 66.

Just as the main microprocessor determines the speed and power of a computer, the CCD chip determines the detail and image quality of a digital camera. CCD chips vary in capability, with the sensing capability determined by the number of pixels on the chip; most have more than 40,000 pixels. Each chip design also yields a different pixel size, ranging from 9 to 27 microns, but the actual resolution achieved by a particular size of pixel is also dependent on the focal length of the telescope being used. For photographing the Moon, smaller chips may outperform larger chips.

For astrophotographers, special features of CCD cameras and the accompanying software can help improve results. Some digital cameras have no shutters like those on traditional cameras, but use *frame transfer* to control exposures. *Antiblooming* is a function that protects against image streaking resulting from exposure to bright objects in the field of view. *Tricolor imaging* produces color results by utilizing three separate filters — one each for red, blue, and green — to record exposures on the same chip. Because the colors are preseparated, the results are open to a wide variety of tweaking. Another special feature of the frames shot in a digital format is *merging*, combining images in a seamless mosaic or layering frames to maximize detail.

The main difference between CCD imaging and conventional photography is speed. CCDs are faster, capturing more information in less time, typically achieving good results with smaller telescopes and smaller apertures. Conventional photography is better for wider fields of view; CCDs excel at high resolutions. Especially for shots of the planets or the Moon, a CCD can capture high resolution shots in very short exposure times. But because CCDs are limited to relatively narrow fields of view, using them to photograph eclipses or other shots of the Moon within a single frame is not feasible.

To adapt a telescope for CCD imaging of the Moon, focal length adapters may be required. Just as with conventional cameras, a neutral density filter or a dark blue filter may help reduce brightness.

Digital cameras produced for the consumer market may also be used for shooting the Moon. However, adapters for telescopes or the right telephoto lenses may not yet be available for all models.

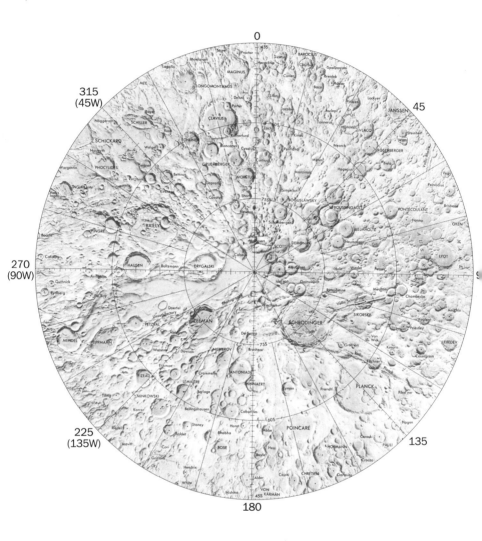

SOUTH POLAR REGION

Stereographic projection of polar areas from 45° to 90°.

(Lunar maps on pages 70 to 74 courtesy of NASA, prepared and published by the Defense Mapping Agency Aerospace Center, St. Louis, Missouri)

SCALE 1:10,000,000 at 34° North and South latitudes.

70

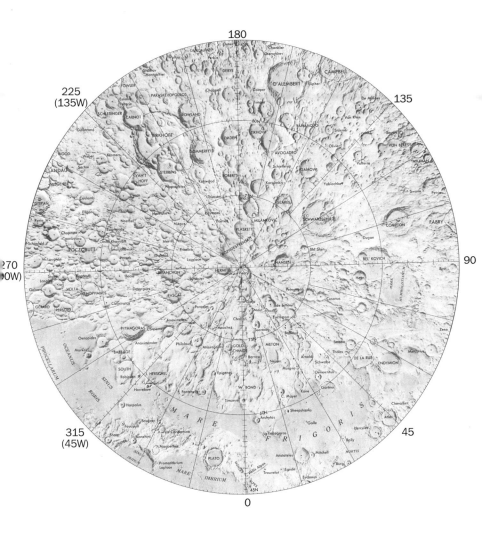

NORTH POLAR REGION

The maps on pages 72–74 are a mercator projection of lunar areas from 45° North to 45° South on the front side, including added area visible during extreme librations. The surface features were interpreted from photographs recorded during Lunar Orbiter Missions I–V.

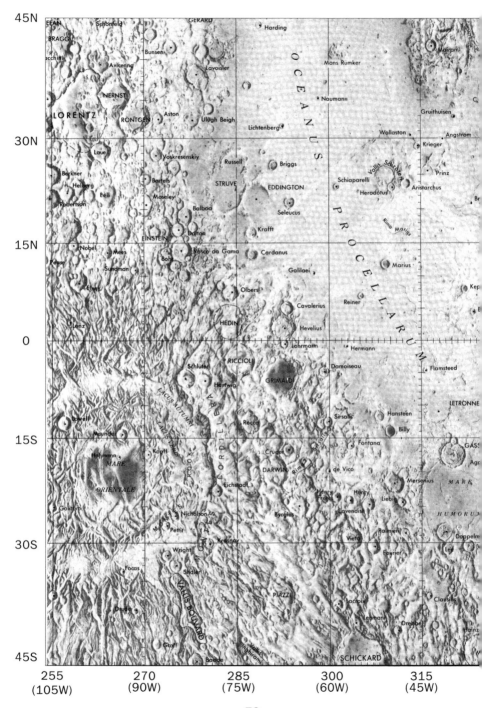

45N

EEAN
BRAGG
acchini
LORENTZ
Schönfeld
Bunsen
Avicenna
NERNST
RÖNTGEN
Aston
Ulugh Beigh
GERARD
Lavoisier
Harding
Lichtenberg
Mons Rumker
Naumann
OCEANUS
Wollaston
Mairan
Gruithuisen
Angstrom

30N

Laue
Berkner
Helberg
Bell
Robertson
Voskresenskiy
Bartels
Moseley
Balboa
Russell
STRUVE
EDDINGTON
Seleucus
Briggs
Schiaparelli
Herodotus
Krieger
Vallis Schröteri
Aristarchus
Prinz
PROCELLARUM

15N

Nobel
Pease
Sundman
Mees
Elvey
Lenz
EINSTEIN
Bohr
Dalton
Vasco da Gama
Olbers
Krafft
Cardanus
Galilaei
Marius
Reiner
Cavalerius
Hevelius
Rima Marius
Kep

0

HEDIN
Lohrmann
Hermann
Flamsteed

Schlüter
RICCIOLI
Hartwig
Rocca
GRIMALDI
Damoiseau
Sirsalis
Hansteen
Billy
LETRONNE

15S

owell
Maunde
Kopff
CORDILLERA
LACUS AUTUMNI
LACUS VERIS
Eichstadt
Nicholson
Pettit
Crüger
DARWIN
Rima Sirsalis
de Vico
Fontana
Henry Frères
Henry
Cavendish
Mersenius
Liebig
Palmieri
GASS
Ago
MARE
HUMORUM
Doppelm

MARE
ORIENTALE
Hohmann
Golitsyn
MONTES ROOK
Krasnov
Wright
Byrgius
Vieta
Fourier
Lee

30S

Focas
MONTES ROOK
Skalar
Drude
VALLIS BOUVARD
Graff
Baude
PIAZZI
Lacroix
Legmann
Drebbel
Clausius
Hainz
SCHICKARD

255
(105W)

270
(90W)

285
(75W)

300
(60W)

315
(45W)

5N

SINUS IRIDUM

Promontorium Helicon
Heraclides
Le Verrier   Kirch

M A R E

Herschel
Heis   Carlini
Delisle

I M B R I U M   Archimedes

0N

Diophantus
Mons La
Hire
Timocharis

Lambert

Euler

yley

Pytheas   Wallace

T. Mayer   MONTES

CARPATUS

5N

Milichius   Copernicus   Stadius

MARE
INSULARUM

icke

Kunowsky   Reinhold
Gambart   Sommering

0

Lansberg   Mosting

Apollo 12   Apollo 14   Flammarion
Fra Mauro

Euclides   MARE   Lalande
Bonpland   Parry   PTOLEMAEUS

COGNITUM   Guericke   Palisa
Herigonius   Davy

Darney   ALPHONSUS

5S

NDI   Lassell   Alpetragius

harchides   Lubiniezky

Bullialdus   MARE   Nicollet   Birt
König   NUBIUM

Hippalus   Kies
Campanus   Hesiodus
Mercator

0S   ello   PITATUS

PALUS   EPIDEMIARUM   Weiss
Capuanus   Cichus
Wurzelbauer   Gauricus
Haidinger   Heinsius
Epimenides
Lagalla   WILHELM   TYCHO

5S   E   E

Piazzi Smyth

Cassini   Alexander
Calippus
Theaetetus

Aristillus

Autolycus   M A R E

PALUS
PUTREDINIS   Fresnel
Apollo 15   S E R E N I T A T I S

Bessel
Canon   Sulpicius Gallus

Marco Polo   Menelaus
MARE   Plinius
Eratosthenes   Manilius
SINUS   Jansen
VAPORUM   Ross
Boscovich   Julius   Maclear
Caesar
Ukert   Rima Hyginus   Rima
Bode   Murchison   Arago   Lamont
Pallas   Trisnecker
Agrippa   Dionysius   Maskelyne
Godin   Ritter   Sabine   Apollo 11

Rhaeticus   Theon Sr.
Theon Jr.   Delambre
Horrocks   Saunder   Taylor   Hypatia
HIPPARCHUS   Alfraganus   Torricelli
Hind   Zollner   Isidorus
Halley   Andel   Apollo 16   Theophilus
Ritchey   Descartes   Kant   Madler
ALBATEGNIUS   Abulfeda   Cyrillus
Vogel   Tacitus
Argelander   Airy   Beaumont   NECTA
Arzachel   Geber   Almanon   Catharina
Faye   Donati   Abenezra   Azophi
Thebit   Delaunay   Fermat
La Caille   Blanchinus   Krusenstein   Sacrobosco   Polybius
PURBACH   Apianus   Pontanus   Altai
REGIOMONTANUS   Werner   Wilkins
Aliacensis   Rothmann
DESLANDRES   Goodacre   Zagut   Lindenau
Hell   WALTER   Gemma   Rabbi Levi
Ball   Lexell   Frisius   Riccius
Furnelius   Bushinq
Miller   Buch
Huggins   STOFLER   Faraday   MAUROLYCUS   Nicolai
Pictet   Nasireddin   BAROCIUS
Saussure

Budaxus   LACUS   MORTIS
Plana   Mason
Grove
Alexander   LACUS
SOMNIORE
Daniell
POSIDONIUS
Chacornac
Le Monnier   Rome
Littrow
Apollo 17   Dawes   Vitruvius
Plinius

M A R E
Sinas

TRANQUIL

Apollo 11   Cen

MA
Abulfeda   Cyrillus

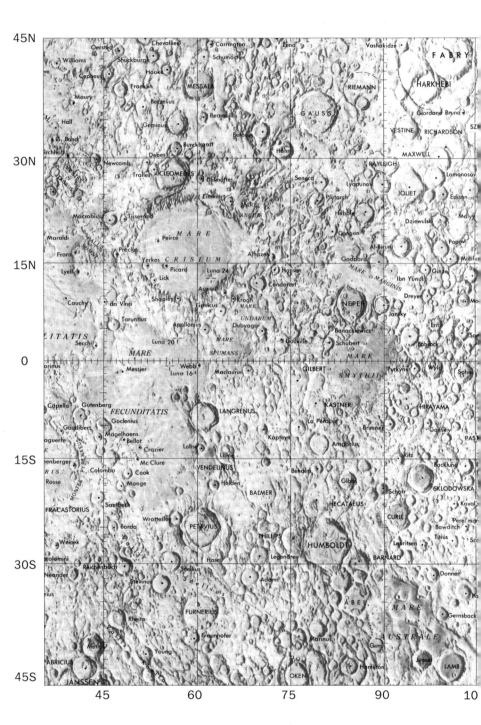

45N

Oersted · Chevallier · Carrington · Zeno · Vashakidze ·

Williams · Shuckburgh · Schumacher · F A B R Y

Cepheus · Hooke · RIEMANN · HARKHEBI

Maury · Franklin · MESSALA · Giordano Bruno ·

Berzelius · VESTINE · RICHARDSON · SZ

Hall · Geminus · Bernouilli · G A U S S ·

G. Bond · Berosus · MAXWELL

irchhoff · Debes · Burckhardt · Hahn · 30N

Newcomb · RAYLEIGH · Lomonosov

Macrobius · Tralles · CLEOMEDES · Delmotte · Senaca · Lyapunov · JOLIET · Edison

Eimmart · Plutarch · Dziewulski · Malyy

Maraldi · Tisserand · Proclus · Peirce · M A R E · Hubble · Canhon · Popov ·

Franz · Yerkes · C R I S I U M · Alhazen · Goddard · Al-Biruni · Möbius · 15N

Lyell · Picard · Luna 24 · Hansen · M A R E · Ibn Yunus · Ginzel

Cauchy · da Vinci · Lick · Auzout · Condorcet · Dreyer · Mo

Taruntius · Shapley · Firmicus · Krogh · MARE · Jansky · Err ·

Apollonius · Dubyago · UNDARUM · Banachiewicz · Babcock · Sa

Secchi · Luna 20 · MARE · Trouville · Schubert · MARE · 0

orinus · Messier · Luna 16 · Maclaurin · GILBERT · Pyrkyne · Wyld · Sgha

Capella · Gutenberg · FECUNDITATIS · LANGRENUS · KÄSTNER · HIRAYAMA

Gaudibert · Goclenius · La Pérouse · Brunner · Gansky · PAS

aguerre · Magelhaens · Bellot · Kapteyn · Ansgarius

enberger · Crazier · Lohse · Lame · VENDELINUS · Ritz · Backlund · 15S

RIS · Colombo · McClure · Behaim · Gibbs · Schorr · SKLODOWSKA

Rosse · Cook · Monge · Holden · BALMER · Koval

FRACASTORIUS · Santbech · Wrottesley · HECATAEUS · CURIE · Perel'man · Bowditch

Weinek · Borda · PETAVIUS · PHILLIPS · Lauritsen · Titius · Scc

scalomini · Hase · Legendre · HUMBOLDT · BARNARD · 30S

Neander · Reichenbach · Snellius · Adams · ÅBEL · Donner

rius · Stevinus · M A R E · Pa

Rheita · FURNERIUS · Marinus · A U S T R A L E · Gernsback

Metius · Fraunhofer · Gum

ABRICIUS · Young · Hamilton · Jenner · LAMB

JANSSEN · OKEN · 45S

45 · 60 · 75 · 90 · 10

74

LUNAR GAZETTEER

Albategnius (crater diameter: 83 miles) 12° South, 4° East
Aliacensis (crater diameter: 50 miles) 31° South, 5° East
Alphonsus (crater diameter: 73 miles) 13° South, 3° West
Anaxagoras (ray crater; crater diameter 32 miles; ray pattern diameter: 600 miles) 75° North, 10° West
Anaximander (crater diameter: 55 miles) 66° North, 48° West
Anaximines (crater diameter: 49 miles) 75° North, 45° West
Archimedes (crater diameter: 50 miles) 30° North, 4° West
Aristillus (ray crater; crater diameter: 35 miles; ray pattern diameter: 400 miles) 34° North, 1° East
Aristoteles (crater diameter: 55 miles) 50° North, 18° East
Arzachel (crater diameter: 61 miles) 18° South, 2° West
Atlas (crater diameter: 54 miles) 47° North, 44° East
Bailly (crater diameter: 184 miles) 66° South, 65° East
Barocius (crater diameter: 54 miles) 45° South, 17° East
Berosus (crater diameter: 45 miles) 33° North, 70° East
Blancanus (crater diameter: 72 miles) 64° South, 21° West
Boguslawsky (crater diameter: 61 miles) 75° South, 45° East
Byrgius (crater diameter: 51 miles) 25° South, 65° West
Casatus (crater diameter: 59 miles) 75° South, 35° West
Catherina (crater diameter: 63 miles) 18° South, 24° East
Clairaut (crater diameter: 47 miles) 48° South, 14° East
Clavius (crater diameter: 144 miles) 58° South, 14° West
Colombo (crater diameter: 46 miles) 15° South, 46° East
Condorcet (crater diameter: 49 miles) 12° North, 70° East
Copernicus (ray crater; crater diameter: 57 miles; ray pattern diameter: 750 miles) 10° North, 20° West
Cuvier (crater diameter: 48 miles) 50° South, 10° East
Cyrillus (crater diameter: 58 miles) 13° South, 24° East

LUNAR FEATURE TERMS

mare	lava plain	*rima*	crack, rille
mons	mountain	*rupes*	cliff
montes	mountain range	*sinus*	bay
palus	dark plain, resembling a swamp	*vallis*	valley

Endymion (crater diameter: 77 miles) 55° North, 55° East
Fabricus (crater diameter: 48 miles) 43° South, 42° East
Faraday (crater diameter: 45 miles) 42° South, 8° East
Fracastorius (crater diameter: 75 miles) 21° South, 33° East
Furnerius (crater diameter: 81 miles) 36° South, 60° East
Gassendi (crater diameter: 69 miles) 18° South, 40° West
Gemma Frisius (crater diameter: 55 miles) 34° South, 14° East
Geminus (crater diameter: 54 miles) 35° North, 57° East
Grimaldi (crater diameter: 127 miles) 6° South, 68° West
Gruemberger (crater diameter: 58 miles) 68° South, 10° West
Hainzel (crater diameter: 46 miles) 41° South, 34° West
Hevel (crater diameter: 69 miles) 2° North, 67° West
Hipparchus (crater diameter: 95 miles) 6° South, 5° East
Hommel (crater diameter: 75 miles) 54° South, 33° East
Humboldt (crater diameter: 130 miles) 27° South, 81° East
Inghirami (crater diameter: 57 miles) 48° South, 70° West
Kepler (ray crater; crater diameter: 20 miles; ray pattern diameter: 400 miles)
    8° North, 38° West
Kircher (crater diameter: 48 miles) 67° South, 45° West
Lacus Somniorum 37° North, 35° East
Lacus Mortis 44° North, 27° East
Langrenus (ray crater; crater diameter: 82 miles; ray pattern diameter: 950
    miles) 9° South, 61° East
Legendre (crater diameter: 46 miles) 29° South, 70° East
Letronne (crater diameter: 73 miles) 10° South, 43° West
Licetus (crater diameter: 47 miles) 47° South, 6° East
Longomontanus (crater diameter: 92 miles) 50° South, 21° West
Maginus (crater diameter: 116 miles) 50° South, 6° West
Manzinus (crater diameter: 60 miles) 68° South, 25° East
Mare Anguis 23° North, 69° East
Mare Australe 50° South, 80° East
Mare Crisium 18° North, 58° East
Mare Fecunditatis 4° South, 51° East
Mare Frigoris 55° North, 0° East
Mare Humboldtianum 55° North, 75° East
Mare Humorum 23° South, 38° West
Mare Imbrium 36° North, 16° West
Mare Marginis 13° North, 87° East

Mare Nectaris 14° South, 34° East
Mare Nubium 19° South, 14° West
Mare Orientale 19° South, 95° West
Mare Serenitatis 30° North, 17° East
Mare Smythii 3° South, 80° East
Mare Spumans 1° North, 65° East
Mare Tranquillitatis 9° North, 30° East
Mare Vaporum 14° North, 5° East
Maurolycus (crater diameter: 72 miles) 42° South, 14° East
Mersenius (crater diameter: 51 miles) 21° South, 49° West
Metius (crater diameter: 54 miles) 40° South, 44° East
Mons La Hire 28° North, 25° West
Mons Piton 41° North, 1° West
Montes Apenninus 20° North, 2° West
Montes Carpatus 15° North, 24° West
Montes Caucasus 36° North, 8° East
Montes Cordillera 27° North, 85° West
Montes Haemus 16° North, 14° East
Montes Jura 46° North, 38° West
Montes Pyrenaei 14° South, 41° East
Montes Riphaeus 6° South, 26° West
Montes Taurus 28° North, 35° East
Moretus (crater diameter: 73 miles) 70° South, 8° West
Mutus (crater diameter: 47 miles) 63° South, 30° East
Neper (crater diameter: 75 miles) 7° North, 83° East
Newton (crater diameter: 85 miles) 78° South, 20° West
Oceanus Procellarum 10° North, 47° West
Olbers (ray crater; crater diameter: 42 miles; ray pattern diameter: 500 miles) 7° North, 78° West
Orontius (crater diameter: 74 miles) 40° South, 4° West
Palus Epidemiarum 31° South, 26° West
Palus Nebularum 38° North, 1° East
Palus Putredinis 27° North, 1° West
Palus Somnii 15° North, 46° East
Petavius (crater diameter: 110 miles) 25° South, 61° East
Phocylides (crater diameter: 75 miles) 54° South, 58° West
Piccolomini (crater diameter: 56 miles) 30° South, 32° East
Pitiscus (crater diameter: 51 miles) 51° South, 31° East

Plato (crater diameter: 63 miles) 51° North, 9° West

Pontecoulant (crater diameter: 60 miles) 69° South, 65° East

Posidonius (crater diameter: 63 miles) 32° North, 30° East

Proclus (ray crater; crater diameter: 19 miles; ray pattern diameter: 400 miles) 16° North, 47° East

Ptolemaeus (crater diameter: 93 miles) 14° South, 3° West

Purbach (crater diameter: 77 miles) 25° South, 2° West

Pythagoras (crater diameter: 80 miles) 65° North, 65° West

Riccioli (crater diameter: 99 miles) 3° South, 75° West

Rosenberger (crater diameter: 61 miles) 55° South, 43° East

Rupes Philolaus 68° North, 25° West

Rupes Recta 22° South, 8° West

Sacrobosco (crater diameter: 60 miles) 24° South, 17° East

Scheiner (crater diameter: 71 miles) 60° South, 28° West

Schickard (crater diameter: 134 miles) 44° South, 54° West

Schiller (crater diameter: 112 miles) 52° South, 39° West

Schomberger (crater diameter: 52 miles) 76° South, 30° East

Sinus Iridum 45° North, 32° West

Sinus Aestuum 12° North, 9° West

Sinus Medii 0°, 0°

Sinus Roris 54° South, 46° West

Snellius (crater diameter: 50 miles) 29° South, 56° East

Stevinus (crater diameter: 46 miles) 33° South, 54° East

Stoflerus (crater diameter: 84 miles) 41° South, 6° East

Strabo (ray crater; crater diameter: 34 miles; ray pattern diamter: 400 miles) 62° North, 55° East

Theophilus (ray crater; crater diameter: 64 miles; ray pattern diameter: 675 miles) 12° South, 26° East

Tycho (ray crater; crater diameter: 54 miles; ray pattern diameter: 1900 miles) 43° South, 11° West

Vallis Alpes 49° North, 2° East

Vallis Rheita 40° South, 48° East

Vendelinus (crater diameter: 94 miles) 16° South, 62° East

Vieta (crater diameter: 53 miles) 29° South, 57° West

Vlacq (crater diameter: 56 miles) 53° South, 39° East

Walter (crater diameter: 82 miles) 33° South, 1° East

Wargentin (crater diameter: 53 miles) 50° South, 60° West

Wilhelm I (crater diameter: 64 miles) 43° South, 20° West

Wurzelbauer (crater diameter: 54 miles) 34° South, 16° West

EARLY MOON WATCHERS

The Moon is not only the most visible symbol in the sky, it has long been the object of study by humans as a measure of the passage of time. Early cultures, however, lacking an understanding of the heliocentric (Sun-centered) nature of the solar system, generally confined their studies to the support of religious beliefs. The first real development in scientific reasoning came from Nicolas Copernicus (1473–1543), who created a theory about the revolution of the Moon around the Earth and the planets around the Sun.

Tycho Brahe, a Danish astronomer (1546–1601), believed in a geocentric (Earth-centered) theory of the universe, but was responsible for developing accurate measurements of the lunar orbit, including the slight variations created from the effect of the Sun's gravity. Johannes Kepler (1571–1630) believed, like Copernicus, that the Sun was the center of the solar system, and made pertinent observations on the Moon's orbit, the lunar effect on tides, and the nature of the lunar surface.

Kepler and Galileo Galilei (1564–1642) both contributed to early lunar research by creating and refining observations of the Moon. Along with the widespread adoption of the telescope by astronomers and scientists in the seventeenth century came the first detailed lunar maps. Johannes Hevelius (1611–1687) was an astronomer who specialized in the study of the lunar surface, ultimately publishing a description of the Moon that included the first standardized names for lunar features. The Hevelius nomenclature has not survived the test of time, being replaced by that of another astronomer from that era, Giovanni Riccioli (1598–1671). Riccioli's lists of lunar features included names of prominent landmarks, craters, and seas, many of which are still in use.

The first map of the Moon based on a system of coordinates was created by Johann Mayer (1723–1762) in 1750. The first lunar landscapes that included measurements were published in 1791 by Johann Schroeter (1745–1816). In 1835, exaggerated reports of the astronomer Sir John Herschel (1792–1871) and his discoveries about the Moon were published in a popular newspaper in New York City.

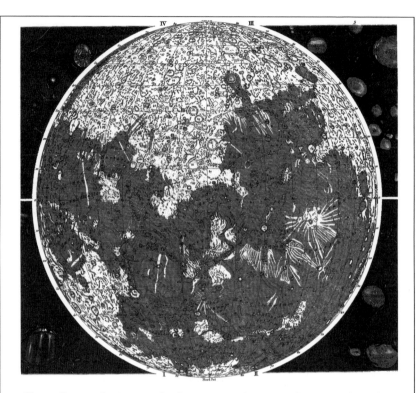

From the earliest use of telescopes, observers have produced maps of the Moon's surface. This line engraving is from *The Iconographic Encyclopedia of Science, Literature and Art*, published in 1851 (R. Garrigue, New York), and reproduces a map created in 1834 by Wilhelm Beer and J. H. von Mädler, German astronomers credited with the first systematic use of names for surface features.

These stories included supposed sightings of lunar animals, but were written without the knowledge or consent of Herschel.

Many other astronomers contributed to the growing study of the Moon, taking advantage of developments in technology and science. The first photograph of the Moon was produced on March 23, 1840, by John Draper (1811–1882). By the end of the nineteenth century, several major observatories published books of photographs of the Moon, including Lick Observatory and the Paris Observatory.

The invention and perfection of rocket-powered flight in the early part of the 20th century was the first step towards a close-up examination of the Moon. Although the scientific theory about space flight to the Moon was already well developed by the 1950s, it took a U.S.-U.S.S.R. "space race" to create the final impetus for the first unpiloted and piloted expeditions to get there.

Hundreds of pounds of moon rocks, close-up human study, and huge quantities of remote sensing data have resulted from the exploration of the Moon. This information has been useful in developing new theories about the solar system and creation of the planets, and has also proven the potential of the Moon for providing huge quantities of valuable materials useful in the further exploration of space.

With a new series of lunar and planetary probes in the 1990s, interest has increased in the exploration of the Moon. Missions are now being discussed and planned for expanded orbital and surface activity. In the new model of space exploration, such missions are increasingly designed as low budget, efficient probes that can be created and flown within a few years of conception. One preliminary finding from the *Clementine* spacecraft, flown in 1994, indicated the possible presence of ice in deep pockets permanently shielded from sunlight and heat. In 1998, a more thorough probe conducted by the *Lunar Prospector* confirmed this finding, setting off a wave of speculation about the potential usefulness of this commodity for future manned Moon bases.

THE FAR SIDE

In 1994, the spacecraft Clementine mapped the entire surface of the Moon using special cameras. This picture of the far side is a mosiac made from thousands of small images that recorded the albedo of surface features.

(Image processing by the U.S. Geological Survey, Flagstaff, Arizona.)

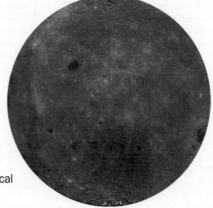

LAGRANGE POINTS

The gravitational physics of orbiting bodies produces a unique condition. Where there are examples of two large celestial bodies orbiting in relation to each other, as in the case of the Earth orbiting around the Sun, five specific points in the orbital patterns have the effect of cancelling the gravitational and centrifugal pull of the bodies. These points are called Lagrange points, after their discovery by Joseph Louis Lagrange, a French mathematician, in 1772.

Lagrange points are potentially important spots for the future, because spacecraft, space stations, or permanent space colonies could remain in stable orbits at these locations without the need for constant fuel expenditure to maintain position.

The Lagrange points in the Earth-Moon system are also affected by additional forces from the Sun. In order to remain unaffected by these forces, objects would have to be placed into elliptical orbits around central points defined by L-4 or L-5.

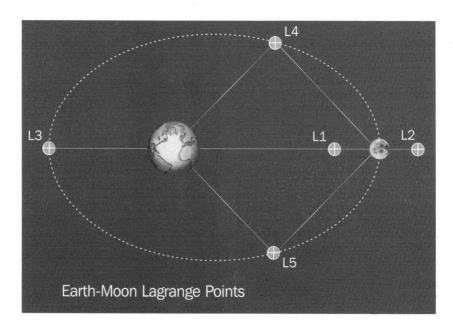

Earth-Moon Lagrange Points

UNPILOTED MOON
MISSION CHRONOLOGY

SPACECRAFT	DATE	MISSION/ACCOMPLISHMENTS • LANDING SITE
Pioneer 1 USA	October 11, 1958	flyby at 71,300 miles
Pioneer 3 USA	December 6, 1958	flyby at 66,654 miles
Luna 1 USSR	September 12, 1959	landed on lunar surface
Ranger 6 USA	February, 1964	cameras malfunctioned
Ranger 7 USA	July 28, 1964	first pre-impact photos • *Mare Nubium, lat 10.6° N, long 24.7° W*
Ranger 8 USA	February 17, 1965	transmitted photographs • *Mare Tranquillitatis, lat 2.6° N, long 24.7° W*
Ranger 9 USA	March 21, 1965	transmitted photographs • *interior of Alphonsus, lat 12.9° S, long 2.4° W*
Luna 5 USSR	May, 1965	unsuccessful soft landing
Zond 3 USSR	July 18, 1965	orbited, transmitted photos (far side and western limb, altitude 9,960 km to 11,570 km)
Luna 7 USSR	October, 1965	unsuccessful soft landing
Luna 8 USSR	December, 1965	unsuccessful soft landing
Luna 9 USSR	January 31, 1966	landed, transmitted photos • *Oceanus Procellarum, lat 7.1° N, long 65.4° W*
Luna 10 USSR	March 31, 1966	orbited, gamma-ray sensing (perilune 350 km, apolune 1,015 km)
Surveyor 1 USA	May 30, 1966	landed, transmitted first color photos, data • *Flamsteed P, lat 2.5° S, long 43.2° W*
Lunar Orbiter USA	August 10, 1966	orbited, transmitted photos, data (inclination 12°, perilune 190–40km, apolune 1,865–1,815 km)
Luna 11 USSR	August 24, 1966	orbited (perilune 165 km, apolune 1,195 km)
Surveyor 2 USA	September, 1966	unsuccessful soft landing
Luna 12 USSR	October 22, 1966	orbited, transmitted data, photos (perilune 100 km, apolune 1,740 km)

CONTINUED

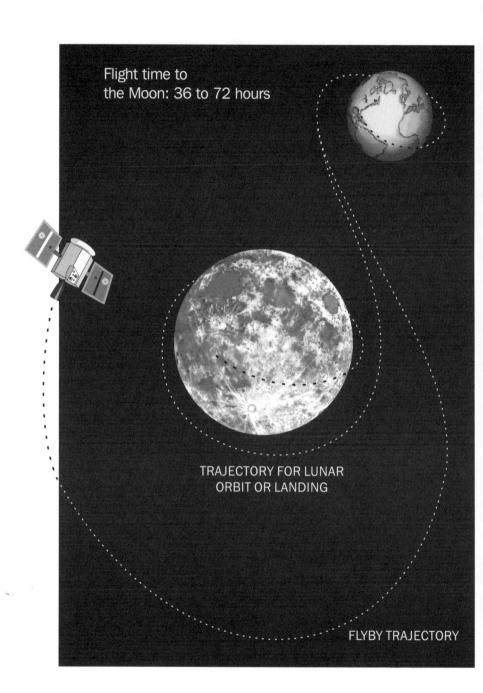

Flight time to
the Moon: 36 to 72 hours

TRAJECTORY FOR LUNAR
ORBIT OR LANDING

FLYBY TRAJECTORY

Lunar Orbiter 2 USA	November 7, 1966	orbited, transmitted photos, data (inclination 12°, perilune 50 km, apolune 1,855 km)
Luna 13 USSR	December 21, 1966	landed, transmitted photos, used mechanical soil sampler
	• *Oceanus Procellarum, lat 18.9° N, long 62.1 W*	
Lunar Orbiter 3 USA	February 4, 1967	orbited, transmitted photos, data (inclination 21°, perilune 55 km, apolune 1,845 km)
Surveyor 3 USA	April 17, 1967	landed, tested lunar soil, transmitted data
	• *Oceanus Procellarum, lat 3.2° S, long 23.4° W*	
Lunar Orbiter 4 USA	May 4, 1967	orbited, remote orbit change (inclination 85°, perilune 2,705 km, apolune 6,115 km)
Surveyor 4 USA	July 14, 1967	contact lost
Explorer 35 USA	July 19, 1967	orbited, magnetic fields (transmitted data until 2/72) (perilune 830 km, apolune 7,650 km)
Lunar Orbiter 5 USA	August 2, 1967	orbited, transmitted data (inclination 85°, perilune 195–100 km, apolune 6,065–1,500 km)
Surveyor 5 USA	September 8, 1967	landed, soil experiments, transmitted data
	• *Mare Tranquillitatis, lat 1.4° N, long 23.1° E*	
Surveyor 6 USA	November 7, 1967	landed, transmitted photos, data
	• *Sinus Medii, lat 0.5° N, long 1.5° W*	
Surveyor 7 USA	January 6, 1968	landed, tested soil, transmitted photos, data
	• *flank of Tycho, lat 40.9° S, long 11.5° W*	
Luna 14 USSR	April 1968	orbited
Zond 5 USSR	September 14, 1968	orbited
Zond 6 USSR	November 10, 1968	flyby, returned film to Earth (altitude about 3,300 km)
Luna 15 USSR	July 1969	landed
Zond 7 USSR	August, 1969	flyby, returned film to Earth (western limb and southern far side, altitude about 2,200–10,000 km)

Luna 16 USSR	September 12, 1970	landed, returned soil samples to Earth
		• *Mare Fecunditatis, lat 0.7° S, long 56.3° E*
Zond 8 USSR	October 1970	flyby, returned film to Earth (altitude about 1,120 km)
Luna 17 USSR	November 10, 1970	landed, used remote vehicle (Lunokhod 1), soil tests, transmitted TV signals
		• *Sinus Iridum, lat 38.3° N, long 35.0° W*
Luna 18 USSR	September 1971	unsuccessful soft landing
Luna 19 USSR	September 28, 1971	orbited, remote sampling (perilune 140–77 km, apolune 140–385 km)
Luna 20 USSR	February 1972	landed, returned soil samples to Earth
		• *Crisium basin rim, lat 3.5° N, long 56.5° E*
Luna 21 USSR	January 8, 1973	landed, remote vehicle, returned to Earth with soil samples
		• *Mare fill of Le Monnier, lat 25.8° N, long 30.5° E*
Luna 22 USSR	May 29, 1974	orbited
Luna 23 USSR	November 1974	damage during landing
Luna 24 USSR	August 9, 1976	landed, returned to Earth with soil samples
		• *Mare Crisium, lat 12.7° N, long 62.2° E*
Muses A JAPAN	January, 1990	orbited
Galileo USA	October 18, 1990	fly-by, remote sampling, photography
Clementine USA	January 25, 1994	orbited, remote measurement, mapping
Lunar Prospector USA	January 6, 1998	orbited, remote sampling, mapping

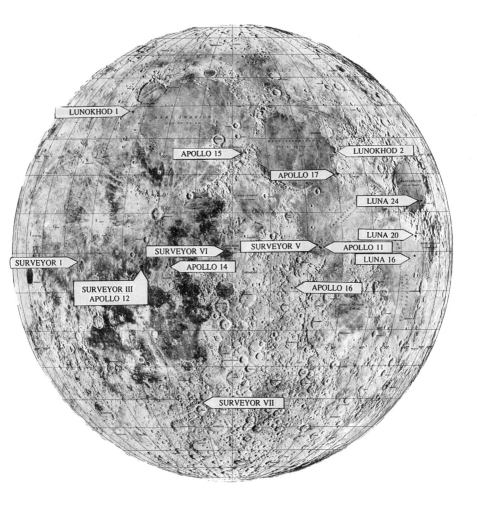

LUNAR LANDING SITE CHART

(Map courtesy of the Lunar and Planetary Institute)

THE APOLLO PROGRAM: MEN ON THE MOON

MISSION	DATE OF LAUNCH	DURATION (days, hours, minutes)	CREW

Apollo 7   October 11, 1968   10d 20h 9m   Cunningham, Eisele, Schirra
- *orbital tests around Earth of Apollo command and service module*

Apollo 8   December 21, 1968 6d 3h 0m   Anders, Borman, Lovell
- *first flight to Moon; first orbit around Moon; inclination about 13°, perilune 110 km*

Apollo 9   March 3, 1969   10d 1h 0m   McDivitt, Schweickart, Scott
- *orbital tests around Earth, first flight of complete Apollo spacecraft*

Apollo 10   May 18, 1969   8d 0h 3m   Cernan, Stafford, Young
- *orbital tests around Moon of complete Apollo spacecraft; partial descent to surface by lunar module; inclination about 1°, perilune 110–15 km)*

Apollo 11   July 16, 1969   8d 3h 18m   Aldrin, Armstrong, Collins
- *first Moon landing; first walk on Moon (Armstrong); Mare Tranquillitatis, Statio Tranquillitatis, lat 0.7° N, long 23.4° E*

Apollo 12   November 12, 1969 10d 4h 36m Bean, Conrad, Gordon
- *landing; surface exploration; Oceanus Procellarum, lat 3.2° S, long 23.4° W*

Apollo 13   April 11, 1970   5d 0h 1m   Haise, Lovell, Swigert
- *flyby; mission aborted in 3rd day*

Apollo 14   January 31, 1971   9d 0h 1m   Mitchell, Roosa, Shepard
- *landing; surface exploration; Fra Mauro highlands, lat 3.7° S, long 17.5° W*

Apollo 15   July 26, 1971   12d 7h 11m   Irwin, Scott, Worden
- *first use of Lunar Rover; first continuous color TV broadcast of moonwalk; extensive scientific study of lunar surface; Palus Putredinis, Apenninus-Hadley region, lat 26.1° N, long 3.7° E*

Apollo 16   April 16, 1972   11d 1h 51m Duke, Mattingly, Young
- *landing; surface exploration; Descartes highlands, lat 9.0° S, long 15.5° E*

Apollo 17   December 7, 1972   12d 13h 1m Cernan, Evans, Schmitt
- *landing; first geological study of lunar surface (Schmitt); Taurus-Littrow valley, lat 20.2° N, long 30.8° E*

THE CREATION
OF THE MOON

Despite an increasing depth of knowledge about the composition of the Moon, scientists have as yet no provable theory about its origin. One general agreement, however, is that the Moon was formed at least 4.5 billion years ago during the creation of the solar system. One theory suggests that it was formed near to, but separate from the Earth; another suggests that it was formed some distance away but was pulled into orbit around the Earth by gravitational attraction. One of the most recent proposals theorizes that a large object such as an asteroid impacted the Earth, throwing off enough vapor and material to form the Moon. During the first phase of the Moon's existence, forces caused by the cooling molten material produced many of the existing features. The largest visible features on the surface were created by violent collisions with asteroids that may have been as large as the state of Delaware. The interior of the Moon is believed to have originally been more solid, with melting caused by radioactive decay. The last violent surface activity on the Moon probably happened more than 3 billion years ago, with volcanic-like eruptions and flooding of lava across the surface.

MOONQUAKES

Much is still unknown about the interior of the Moon, but scientists believe that there is a core of molten or partly molten material. There are different layers in the structure of the moon, and a constant shifting produces tremors similar to earthquakes. These moonquakes are usually very weak — many of them release no more energy than a firecracker. Some moonquakes are caused by the impact of meteorites on the surface. Others occur at regular intervals during a lunar cycle, suggesting that gravitational forces from the Earth similar to ocean tides are causing movement within the body.

MOON ROCKS

Examination of the rocks brought back by the Apollo astronauts is still underway. Scientists have discovered many interesting features about the composition and origin of the Moon from these rocks, most of them formed from cooling lava and therefore igneous in nature. Some of the rocks are similar to basalt which is found on Earth. Samples of this basalt were collected in low areas of the lunar surface that are observed as maria from the Earth.

Rocks from higher regions of the Moon are also igneous, and are referred to as gabbro, norite, and anorthosite, similar to rocks of the same names on Earth. Although moon rocks have some characteristics similar to Earth rocks, they are recognizably different because of the complete lack of water and the effects that water has on minerals in rocks. Moon rocks also exhibit crystals of metallic iron that occur because of the lack of free oxygen. Lunar minerals include feldspar, olivine, pyroxine, ilmenite, plagioclase, and troilite.

HOW OLD IS THE MOON?

The best guess from studying the age of lunar rocks is that the Moon was formed about 4.5 billion years ago. The newest rocks found so far on the Moon (from the dark maria regions) have been dated at 3.1 to 3.8 billion years old.

Material on the surface of the Moon is referred to as regolith or lunar soil, but it has no organic content. Lunar soil forms a layer from 3 to 60 feet (1 to 20 meters) deep on the surface. This layer is composed of rocks and powder but was not formed by eroding forces such as wind or water. Instead, lunar soil was created over a period of billions of years by the continuous bombardment of meteorites. Larger meteorites form visible craters that can be seen from the Earth; smaller, virtually invisible craters are formed by particles of cosmic dust. The smallest craters are only 1/25,000 inch (1/1000 millimeter) in diameter.

MOON CALENDARS

Most ancient civilizations based their calendars on the lunar cycle. The highly visible phase changes of the Moon made observations and accurate date projections relatively easy. There is, however, a built-in problem with this method. Most of the seasonal variations in climate are linked to the solar year, and there is no even number of lunar months which equals one solar year. Therefore calendars based on lunar months are always out of step with the seasons, and extra days or months must be added periodically to make them practical. This procedure is referred to as intercalation.

The ancient Babylonian calendar was based on a lunar month that began when the first crescent Moon was visible. There were 12 months in every year, with months alternating between 30 and 29 days long. This lunar calendar would have been out of phase with the solar year, which is almost 12 days longer, except that the Babylonians added special intermediate (or intercalary) months seven times in every calendar cycle. The extra days in these special months made the cycle come out even (start repeating itself) every 19 years.

The 19-year cycle was also adopted by other cultures, including the Greeks, and is also used in the present-day Jewish calendar. Each of these cycles is the equivalent of 235 lunar months, or lunations.

The Egyptian civilization was heavily influenced by the annual flooding of the Nile River. The time of flooding every year coincided with the rising of the star Sirius close to the Sun, so the Egyptians created a lunar calendar that had its first month begin when the new moon occurred after the rising of Sirius. This lunar calendar featured 12 months, each one 29½ days long. Extra months were included occasionally to keep pace with the solar year. The Egyptians were very advanced in observations of celestial cycles and eventually created a solar calendar to replace the lunar one.

Early Greeks developed and relied on lunar calendars that were organized and maintained by each separate city or town. Beginning about the sixth century B.C., Greek astronomers and mathematicians created more organized lunar calendars based on 19-year cycles, but they eventually switched to the Roman calendar, which was based on the Sun.

The Romans originally used a lunar calendar but reorganized it during the reign of Julius Caesar. Caesar added extra days to the lunar calendar to keep it from getting out of step with the solar year; the new calendar was first used in 45 B.C. This calendar was referred to as the Julian calendar and was widely used in western countries until a further reform was organized in 1582. At that time, the rapid spread of Christianity necessitated the fixation of religious holidays based on the celebration of Easter. Pope Gregory XIII was responsible for this reform, and the new calendar, still in use today, is called the Gregorian calendar. The Gregorian calendar measures a year at 365 days (an average of 365.2425 days if leap year is included) and comes very close to matching the solar cycle. Every 400 years, the total difference between the two is only a few hours. By comparison, the error with the Julian calendar was about three days for the same period.

The Gregorian calendar which we use today is based on a demand from the Christian religion to determine the correct date for Easter every year. Easter is considered the beginning of the Christian calendar and all Christian holidays and special days are figured from that date. Easter is determined by a fixed set of rules that modify the actual lunar cycle in order to keep the date between March 22 and April 25. If the real, observable lunar cycle were used, the date for Easter could vary much more.

In practice, a fixed date for the spring equinox is set at March 21 and Easter is determined by finding the first Sunday following the full moon

CHINESE LUNAR MONTHS

1	Holiday Moon	8	Harvest Moon
2	Budding Moon	9	Chrysanthemum Moon
3	Sleepy Moon	10	Kindly Moon
4	Peony Moon	11	White Moon
5	Dragon Moon	12	Bitter Moon, Moon of
6	Lotus Moon		Offerings
7	Moon of Hungry Ghosts		

that occurs on or just after March 21. This full moon date is not determined from the observable cycle but by special religious tables that vary slightly from the real times. Not by coincidence, these tables are based on the same 19-year lunar cycle which was used by the Babylonian, Greek, and Jewish calendars.

Among those calendars that rely on the Moon, lunar months may run from full moon to full moon or new moon to new moon, depending on tradition. In southern India, the month begins with the new moon; in other parts of the country, the full moon signals the beginning.

The Jewish calendar was first created around the sighting of the crescent moon, which determined the beginning of each of 12 lunar months. At occasional intervals over the years, the twelfth month was repeated to synchronize the calendar with the seasons. Until the fourth century A.D., there were various versions of this calendar in use. Jews in some countries began their year with the month of Nisan in the

RELIGIOUS MONTHS

Lunar months have traditional names in some major religions.

JEWISH		MOSLEM
Tishri	1	Muharram
Heshvan	2	Safar
Kislev	3	Rabi'u'l-Avval
Tevet	4	Rabi'u'th-Thani
Shevat	5	Jumadiyu'l-Avval
Adar	6	Jumadiyu'th-Thani
Nisan	7	Rajab
Iyyar	8	Sha'ban
Sivan	9	Ramadan
Tammuz	10	Shavval
Av	11	Dhi'l-Qa'dih
Elul	12	Dhi'l-Hijjih

HAWAIIAN LUNAR DAYS

day 1   Hilo (first appearance of crescent moon)
2   Hoaka
3   Ku-Kahi
4   Hu-lua
5   Hu-kolu
6   Hu-pau
7   Ole-ku-kahi   FIRST QUARTER MOON
8   Ole-ku-lua
9   Ole-ku-kolu
10   Ole-ku-pau
11   Huna
12   Mohalu
13   Hua
14   Akua   FULL MOON
15   Hoku
16   Mahealani
17   Kulu 18 Laau-ku-kahi
18   Laau-ku-lua
20   Laau-pau
21   Ole-ku-kahi   LAST QUARTER MOON
22   Ole-ku-lua
23   Ole-pau
24   Kaloa-ku-kahi
25   Kaloa-ku-lua
26   Kaloa-pau
27   Kane
28   Lono
29   Mauli ("moon is fainting")
30   Muku ("moon cut off by the Sun")

spring; other countries used the month of Tishri in the fall. The modern Jewish calendar was adopted in the fourth century A.D. and relies on fixed sequences of months based on a 19-year cycle of lunations. This 19-year cycle is the same one that was used by the Babylonians. The new calendar determines the beginning of the year from the time of the new moon in the month of Tishri (occurring in the fall) but uses special rules to keep important religious days from falling on the wrong days. The Jewish year can have 353, 354, or 355 days, with occasional "leap years" of 383, 384, or 385 days. For every 19-year cycle, there are 12 regular years and seven "leap years."

The Moslem calendar is also based on lunar months but uses a cycle of 33 years. There are 12 lunar months in the Moslem calendar, with lengths alternating between 29 and 30 days. The calendar is fixed instead of relying on actual observation of the Moon, but it is only off from the actual lunar cycle by one day every 2,500 years.

In practice, many Moslems base their religious holidays on actual

HAWAIIAN LUNAR MONTHS

1 Moakali'i (also Hui-hui, refers to the Pleiades)
2 Ka'elo ("Drenching Month," refers to Betelgeuse)
3 Kaulua ("Pit of Sacrifice," named for Sirius)
4 Nana
5 Welo
6 Ikiiki ("Warm and sticky")
7 Ka'aona
8 Hinaia-'ele'ele
9 Hili-na-ehu (word for sea-borne mists)
10 Wehewehe
11 Hili-na-ma
12 Ikuwa (also Kauka-Malama, Welehu, Welehu-lua,
    or Welehu II, an extra month used to compensate for
    the rising of the Pleiades when it comes
    late in the year)

NEW GUINEA
LUNAR MONTHS

Mu'o'eng (Rainbow Fish moon)
Mumbla'eng (Black Trevally moon)
Mumbe'eng (Blue-speckled Parrotfish moon)
Nane'eng (Palolo Worm True moon)
Nantun'eng (Palolo Worm Second moon)
Momo'eng (Tiger Shark moon)
Mupi'eng (Flying Fish moon)
Pobape'eng (Rain and Wind moon)
Ubala'eng (Logs and Open Sea moon)

observation but rely on a modified lunar calendar — or western Gregorian calendar — for day-to-day activities. The important dates are figured from the first observation of the crescent moon, and these religious days begin at sunset. The first year of use for the Moslem calendar was 622 A.D., the year Mohammed left Mecca.

Most of India also uses the Gregorian calendar for day-to-day activities, but Hindu religious dates are based on a very old lunar calendar. The traditional Hindu calendar dates back to at least 1000 B.C. It is organized around 12 lunar months of 27 or 28 days each. About every 60 months, an extra month is added to keep it in step with the solar year. Each month is also divided into two parts, corresponding to the waxing and waning phases. In addition, the Hindu calendar also uses a

THE FIRST WORD

The English word "moon" gradually developed from early versions. In Old English, the word was "mona" and in Middle English, "mone" or "moone." In Latin, the word was "mensis," and Greek "men." Gothic was "mena." In old Nordic, the word was "mani" and in old high German, it was "mano."

INDIAN ASTROLOGY

Traditional Indian astrology is based on 27 lunar months, called nakshatra, or mansions. Each begins with the rising of the Moon at a particular point on the ecliptic.

1	Aswini		15	Swati
2	Bharani		16	Visakha
3	Krittika		17	Anuradha
4	Rohini		18	Jyeshtha
5	Mriga		19	Mula
6	Ardra		20	Purva-shadha
7	Punarvasu		21	Uttara-shadha
8	Pushya		22	Stravana
9	Ashlesha		23	Dhanishtha
10	Magha		24	Satataraka
11	Purva-phalguni		25	Purva-bhadrapada
12	Uttara-phalguni		26	Uttara-bhadrapada
13	Hasta		27	Revati
14	Chitra			

complicated system relating Earth's solar orbit with Jupiter's, and uses a unique division of daily time into *vipalas* (0.4 seconds each) and *ghatikas* (24 minutes each).

The oldest lunar calendar currently in use is Chinese. At least one estimate puts the first year of use of the traditional Chinese calendar at 2698 B.C. This calendar is used to determine dates of traditional festivals and special religious days, but it is no longer used for daily or civil activities. It was banned in 1912 when the Republic of China was formed and the traditional name for this calendar system was changed to *Nung Li*, meaning Farmers' Calendar. The Chinese government and most of the country now rely on the western Gregorian calendar for day-to-day activities.

The traditional Chinese calendar is based on 12 lunar months, with each month having either 29 or 30 days. Occasionally a special inter-

FOREIGN MOONS

AFRIKAANS maan
ARABIC kamar or qamar
BASQUE hilargi
BRETON loar
CARIB nu'nû
CATALAN lluna
CEBUANO bulan, buwan
CHECHEN butt
CHINESE yueh
CHINESE (PINYIN) yuèqiú
CHINESE MANDARIN yuèliang
DANISH måne
DUTCH maan
EGYPTIAN pooh
ESKIMO tatkret
ESPERANTO luno
ETHIOPIAN sin (or ilmuqah)
FIJIAN vula
FINNISH kuu
FRENCH lune
GAELIC gealach
GERMAN mond
GREEK selina, mena
GUJARATI chandra
HAITIAN CREOLE lalin
HAUSA wata
HEBREW yaréakn
HINDI-URDU chad

HUNGARIAN hold
HINDUSTANI chandra
ICELANDIC tungli
INDONESIAN bulan
IRISH GAELIC gealach, luan
JAPANESE hyourin, tsuki
JIVARO nantu
KAMBA mwei
KOREAN tal
KURDISH meh, hîv
LAO pa: cha:n
LATIN luna
LATVIAN meness
LITHUANIAN menu
MACEDONIAN mu:n
MALAY bulan
MAYAN luunaa
MONGOLIAN sar
NEPALI candrama, jun
NIUEAN mahina
NORWEGIAN måne
PALAUAN búil
PANJABI cann
PERSIAN mah
PERUVIAN INDIAN sin
PHILIPINO buwán
POLISH ksiezic
PORTUGESE lua

PULAAR lewru
PYGMY pe
ROMANIAN luna
RUNYANKORE okwêzi
RUSSIAN luna
SAMOAN māsina
SANSKRIT mas, chandraḥ
SERBO-CROATIAN mjesec
SLOVAK mesiac, luna
SOMALI dayax
SPANISH luna

SRANAN mun
SWAHILI mwezi
SWEDISH måne
THAI prá-jun
TIBETAN dah-wah
TIV uwer
TURKISH ay, mehtap
WELSH lleuad, looer
YIDDISH levone
YORUBA òṣùpá

AMERICAN SIGN LANGUAGE

The sign for Moon is formed by making the shape of the letter "C" with the right hand and holding it up to the right eye.

NATIVE AMERICAN MOONS

ATAKAPA iti', yil

BILOXI ina

BLACKFOOT natósi, ki'sómm

CHOCTAW hạshi

COCHITI tâ'waṭạ

CHEROKEE nvda

CREE tipiska'wi-pi'sim

DAKOTA SIOUX wi

DELAWARE nipáhum

HAIDA qoñ

HAWAIIAN mahina

HOPI muuyaw

HUPA xatLe wha

INUKTITUT ESKIMO tatkret

JEMEZ p̂â

KAMIA xashá

KILIWA xa?la?

KOASATI nithahasí

KWAKIUTL ᴇmEku'la

MENOMINI típä'kē 'so

MICMAC kisuhs, depkunoosat

MISKITO kati

MUSKOGEE hvréssē

NARRAGANSETT nanepaùshat

NATICK nepauzshad

NAVAJO ooljéé, tł 'ééhona'ái

NEZ PERCE hí-semtuks

OFO i'la

OJIBWAY ne-bah-geeses, dee-bee-kee-zeis

ONONDAGA garáchqua

OSAGE wa-kon̄'-da hon-don

PAPAGO-PIMA marshad

POTAWATOMI tpukisIs

SHOSHONE mea

TAOS p̂aenâ

TLINGIT dîs

ZUNI jáůnanne

mediate month was added to keep the lunar months in line with the solar year. The calendar runs on a 60-year cycle, at which time it begins repeating itself. Inside this big cycle, however, are five smaller cycles, each lasting 12 years. These 12 years are named after animals. The Chinese new year starts with the first new moon that occurs after the Sun has entered the constellation of Aquarius. In practice, New Year's Day in the Chinese calendar can occur from January 20 to February 19.

The traditional culture of the Hawaiian Islands also has a lunar calendar, one based on the rising of stars. Each lunar month is called a *mahina* and begins when the Sun sets on the first day after the new moon. Months are either 29 or 30 days in length and each day in a month is named for the appearance of the Moon on that day or a period of time after a particular phase. The first lunar month begins with the rising of the Pleiades constellation in the east ("Hakali'i" or "Hui-hui" in Hawaiian).

In modern times, some major religions still base calendars on a lunar timetable. Both the Hebrew and Islamic calendar, for example, use the moon as a basis for calculating the passage of time. In the case of Moslem tradition, orthodox or fundamentalist members believe that the beginning of each lunar month cannot begin until the first crescent moon has actually been seen by human eyes. But although the practice of looking for the first crescent moon continues, modern crescent sighters sometimes rely on computer programs, cell phones, and the Internet to maximize their efforts (see "First Sighting," page 43).

AMERICAN MOONS

American colonists brought many European traditions with them when they settled this country. Among those traditions was the naming of full moons. These traditional names were often connected to religious — mostly Christian — dates. In the New World, the naming of full moons was also influenced by the traditions already established in northeastern North America by Native American tribes, mostly Algonquin.

Tribes in other parts of the country often had different names for the moons, usually related to natural changes caused by the seasons. In some cases, twelve distinct names were used but there were tribes who used no more than six names for an entire year. Tribes also shared moon names and sometimes changed the name of a moon from one year to the next. Even among members of a single tribe, scattered groups might sometimes use different names for the same moon or the same name for different moons. The Santee band of the Dakotah Sioux, for example, traditionally called the full moon in September, "Moon When the Horns Are Broken Off," but the same name was used for the full moon in December by the Teton band of the same tribe.

JANUARY *The first full moon after the winter solstice or the first full moon after Yule.*

Colonial American	Winter Moon (also Yule Moon)
Algonquin	Wolf Moon (also Old Moon)
Cherokee	Month of the Cold Moon
Cheyenne	Hoop and Stick Game Moon
Choctaw	Cooking Moon
Dakotah Sioux	Moon of the Terrible
Haida	Younger Moon
Ildefonso	Ice Moon
Klamath	Moon of the Little Finger's Partner
Kutenai	Naktasu Moon (no translation)
Laguna	Lizard Cut Moon
Lakota Sioux	Moon of Frost in the Teepee
Micmac	Frost-Fish Moon
Mohawk	The Big Cold
Natchez	Cold Meal Moon

Nez Perce	Cold Weather Moon
Ojibway	Great Spirit Moon
Osage	Hunger Moon
Oto	The Little Young Bear Comes Down the Tree
Ponca	Snow Thaws
San Juan	Ice Moon
Taos	Man Moon
Tlingit	Goose Moon
Wisham	Her Cold Moon
Zuni	Trees Broken Moon (same as July)

FEBRUARY *The second full moon of the year, associated with the middle of winter.*

Colonial American	Trapper's Moon, Snow Moon, Storm Moon
Algonquin	Snow Moon (also Hunger Moon)
Cherokee	Month of the Bony Moon
Cheyenne	Big Hoop and Stick Game Moon
Choctaw	Little Famine Moon
Dakotah Sioux	Moon of the Raccoon, Moon When Trees Pop
Haida	Elder Moon
Kutenai	Black Bear Moon
Laguna	Yamuni Moon (Yamuni is an edible root)
Lakota Sioux	Moon of the Dark-red Calves
Micmac	Snow-Blinding Moon
Mohawk	Lateness
Natchez	Chestnut Moon
Nez Perce	Budding Time
Ojibway	Sucker Moon
Osage	Light of Day Returns Moon
Oto	Raccoon's Rutting Season
Ponca	Moon When the Ducks Come Back to Hide
San Ildefonso	Wind Moon
San Juan	Coyote Frighten Moon
Taos	Winter Moon
Tewa	Moon When the Coyotes are Frightened
Tlingit	Black Bear Moon
Wisham	Shoulder Moon
Zuni	No Snow on Trails Moon (same as August)

NATIVE AMERICAN SIGN LANGUAGE

Many of the Indian cultures in the western regions of North America shared a common language of hand signs. Using this sign language, one way to signify the word "moon" was as follows. First, the sign was made for the word "night," by extending both hands in front of the body with the palms down and a few inches apart. The right hand is held a little higher than the left hand and both hands are turned inward, crossing over one another. Next, the right hand was used to form the letter "C" by curving the index finger and thumb, the ends about an inch apart, and the other fingers closed. This hand is then raised above the head. The names of specific months could also be signalled by adding another sign to describe the activity traditionally associated with that month.

MARCH *The third full moon of the year.*

Colonial American	Fish Moon (also Worm Moon, Sap Moon, Crow Moon, Lenten Moon, Chaste Moon)
Algonquin	Worm Moon (also Crow Moon, Crust Moon, Sap Moon)
Cherokee	Month of the Windy Moon
Cheyenne	Light Snow Moon (also Dusty Moon)
Choctaw	Big Famine Moon
Dakotah Sioux	Moon When Eyes Are Sore from the Bright Snow
Delaware	Moon when the Juice Drips From the Trees
Haida	Tahet Moon (Tahet is a type of salmon)
Kutenai	Earth Cracks Moon
Laguna	Schamu Moon (Schamu is a local plant)
Lakota Sioux	Moon of Snow-blindness
Micmac	Spring Moon
Mohawk	Much Lateness
Natchez	Deer Moon
Nez Perce	Flower Time
Ojibway	Breaking Up of Snow Shoes Moon
Osage	Just Doing That Moon

Oto	Big Clouds Moon
Ponca	Sore-Eyes Moon
San Ildefonso	All Leaf Split Moon
San Juan	Lizard Moon
Taos	Wind Strong Moon
Tlingit	Moon When Sea Flowers and All Other Things Under the Sea Begin to Flower
Wisham	The Seventh Moon (also Long Days Moon)
Zuni	Little Sandstorm Moon (same as September)

APRIL *The fourth full moon of the year.*

Colonial American	Planter's Moon (also Easter Moon, Pink Moon, Grass Moon, Egg Moon, Seed Moon)
Algonquin	Pink Moon (also Sprouting Grass Moon, Egg Moon, Fish Moon)
Cherokee	Month of the Flower Moon
Cheyenne	Spring Moon
Choctaw	Wildcat Moon
Dakotah Sioux	Moon When Geese Return in Scattered Formations, Moon to Go Paddling
Haida	Ketkakaitash Moon (no translation)
Illinois	Do Nothing Moon
Kutenai	Deep Water Moon
Laguna	Sticky Mud Plant Moon
Lakota Sioux	Moon of Grass Appearing
Micmac	Egg-Laying Moon
Mohawk	Budding Time
Natchez	Strawberries Moon
Nez Perce	Kaket Time (Kaket is an edible root)
Ojibway	Boiling Down of Sap Moon
Osage	Planting Moon
Oto	Little Frogs Croak Moon
Ponca	Rains Moon
San Ildefonso	Leaf Spread Moon
San Juan	Leaf Split Moon
Taos	Ashes Moon
Tlingit	Real Flower Moon
Wisham	The Eighth Moon
Zuni	Great Sandstorm Moon (same as October)

MAY *The fifth full moon of the year.*

Colonial American	Milk Moon (also Mother's Moon, Hare Moon)
Algonquin	Flower Moon (also Corn Planting Moon, Milk Moon)
Cherokee	Month of the Planting Moon
Cheyenne	Time When the Horses Get Fat
Choctaw	Panther Moon
Dakota Sioux	Moon to Plant, Moon When Leaves Are Green
Haida	Salmonberry Bird Moon
Kutenai	Deep Water Moon
Laguna	Loam Plant Moon
Lakota Sioux	Moon of the Shedding Ponies
Micmac	Young Seals Moon
Mohawk	Time of Big Leaf
Natchez	Little Corn Moon
Nez Perce	Kouse Bread Time
Nunamiut Eskimos	Moon When the Ice Goes Out of the Rivers
Ojibway	Budding Moon
Osage	Little Flower Killer Moon
Oto	To Get Ready for Plowing and Planting
Ponca	Summer Begins Moon
San Ildefonso	Planting Moon
San Juan	Leaf Tender Moon
Taos	Corn Planting Moon
Tlingit	Moon When People Know That Everything Is Going to Grow
Wisham	The Ninth Moon
Zuni	Moon No Name Moon (same as November)

JUNE *The sixth full moon of the year, also the full moon closest to the summer solstice.*

Colonial American	Rose Moon (also Stockman's Moon, Strawberry Moon, Honey Moon, Hot Moon, Flower Moon, Dyad Moon)
Algonquin	Strawberry Moon
Cherokee	Month of the Green Corn Moon
Cheyenne	Moon When the Buffalo Bulls are Rutting
Choctaw	Windy Moon
Dakotah Sioux	Moon When June Berries Are Ripe
Haida	Berry Ripening Season Moon
Kutenai	Ripening Strawberries Moon

Laguna	Corn Moon
Lakota Sioux	Moon of Making Fat
Micmac	Leaf-Opening Moon
Mohawk	Ripening Time
Natchez	Watermelons Moon
Nez Perce	Salmon Fishing Time
Ojibway	Strawberry Moon
Osage	Buffalo Pawing Earth Moon
Oto	Hoeing Corn Moon
Ponca	Hot Weather Moon
San Ildefonso	Flower Moon
San Juan	Leaf Dark Moon
Taos	Corn Tassle Appear Moon
Tlingit	Moon of the Salmon
Wisham	Rotten Moon
Zuni	Turning Moon (same as December)

JULY *The seventh full moon of the year.*

Colonial American	Summer Moon (also Buck Moon, Thunder Moon, Hay Moon, Mead Moon)
Algonquin	Buck Moon (also Thunder Moon)
Cherokee	Month of the Ripe Corn Moon
Choctaw	Crane Moon
Dakotah Sioux	Moon When Chokeberries Are Red, Middle of the Summer Moon
Haida	Killer Whale Moon
Kutenai	Ripening Service Berries Moon
Laguna	Corn Tassle Moon
Lakota Sioux	Moon when the Cherries are Ripe
Micmac	Sea-Fowl Shed Feathers
Mohawk	Time of Much Ripening
Natchez	Peaches Moon
Nez Perce	Red Salmon Time
Ojibway	Raspberry Moon
Osage	Buffalo Breeding Moon
Oto	Buffalo Rutting Season Moon
Pima	Moon of the Giant Cactus
Ponca	Middle of Summer Moon
San Ildefonso	Rain Moon
San Juan	Ripe Moon

Taos	Sun House Moon
Tlingit	Moon When Everything Is Born
Wisham	Advance in a Body Moon
Zuni	Trees Broken Moon (same as January)

AUGUST *The eighth full moon of the year.*

Colonial American	Dog Day's Moon (also Woodcutter's Moon, Sturgeon Moon, Green Corn Moon, Grain Moon, Wort Moon)
Algonquin	Sturgeon Moon (also Red Moon, Green Corn Moon)
Cherokee	Moon of the End of the Fruit Moon
Cheyenne	Time When the Cherries are Ripe
Choctaw	Women's Moon
Dakotah Sioux	Moon When All Things Ripen
Haida	Collect Food for Winter Moon
Kutenai	Berries Ripen Even in the Night Moon
Laguna	Yamoni Moon (Yamoni is an immature ear of corn)
Lakota Sioux	Moon When the Cherries Turn Black
Micmac	Young Birds Are Full-Fledged
Mohawk	Time of Freshness
Natchez	Mulberries Moon
Nez Perce	Summer Time
Ojibway	Blueberry Moon
Osage	Yellow Flower Moon
Oto	All the Elk Call Moon
Ponca	Corn Is in Silk Moon
San Ildefonso	Wheat Cut Moon
San Juan	Wheat Cut Moon
Taos	Autumn Moon
Tlingit	Moon When All Kinds of Animals Prepare Their Dens
Wisham	Blackberry Patches Moon
Zuni	No Snow on Trails Moon (same as February)

SEPTEMBER *The ninth full moon of the year, also the full moon closest to the fall equinox.*

Colonial American	Harvest Moon (also Fruit Moon, Dying Grass Moon, Barley Moon)
Algonquin	Harvest Moon
Cherokee	Month of the Nut Moon
Cheyenne	Cool Moon
Choctaw	Mulberry Moon

Dakotah Sioux	Moon When Wild Rice Is Stored for Winter Use
Haida	Salmon Spawning Moon
Kutenai	Ripe Choke Cherries Moon
Laguna	Corn in the Milk Moon
Lakota Sioux	Moon when the Calves Grow Hair (also Moon of the Black Calves and Moon when the Plums are Scarlet)
Micmac	Moose-Calling Moon
Mohawk	Time of Much Freshness
Natchez	The Great Corn Moon
Nez Perce	Spawning Salmon Time
Ojibway	Wild Rice Moon
Osage	Deer Hiding Moon
Oto	Spider Web on the Ground at Dawn Moon
Paiute	Moon Without a Name
Ponca	Moon When the Elk Bellow
San Ildefonso	All Ripe Moon
San Juan	All Ripe Moon
Taos	Leaf Yellow Moon
Tlingit	Small Moon
Wisham	Her Acorns Moon
Zuni	Little Sandstorm Moon (same as March)

OCTOBER *The tenth full moon of the year, also the full moon after the fall equinox and the Harvest Moon.*

Colonial American	Hunter's Moon (also Blood Moon)
Algonquin	Hunter's Moon
Cherokee	Month of the Harvest Moon
Cheyenne	Moon When Water Begins to Freeze on the Edge of the Stream
Choctaw	Blackberry Moon
Dakotah Sioux	Moon When Quilling and Beading Is Done
Haida	Kaganakyash Moon (no translation
Kutenai	Falling River Moon
Laguna	Ripe Corn Moon
Lakota Sioux	Moon of the Changing Season
Micmac	Fat, Tame Animals Moon
Mohawk	Time of Poverty
Natchez	Turkeys Moon
Nez Perce	Falling Leaves Time

Ojibway	Falling of the Leaves Moon
Osage	Deer Breeding Moon
Oto	Deer Rutting Season Moon
Ponca	They Store Food in Caches Moon
San Ildefonso	Harvest Moon
San Juan	Leaf Fall Moon
Taos	Corn Ripe Moon
Tlingit	Big Moon
Wisham	Her Leaves Moon (also Travel in Canoes Moon)
Zuni	Great Sandstorm Moon (same as April)

NOVEMBER *The eleventh full moon of the year.*

Colonial American	Beaver Moon (also Frosty Moon, Snow Moon)
Algonquin	Beaver Moon
Cherokee	Month of the Trading Moon
Cheyenne	Freezing Moon
Choctaw	Sassafras Moon
Dakotah Sioux	Moon When Horns Are Broken Off, Moon When Deer Copulate
Haida	Stomach Moon
Kutenai	Killing Deer Moon
Laguna	Autumn Moon
Lakota Sioux	Moon of the Falling Leaves
Micmac	Tomcod Moon
Mohawk	Time of Much Poverty
Natchez	Bison Moon
Nez Perce	Autumn Time
Ojibway	The Freezing Up Moon
Osage	Coon Breeding Moon
Oto	Every Buck Loses His Horns Moon
Ponca	Beginning of Cold Weather Moon
San Ildefonso	All Gathered Moon
San Juan	All Gathered Moon
Taos	Corn Harvest Moon
Tlingit	Moon When People Have to Shovel Snow Away from Their Doors
Wisham	Her Frost Moon (also Snowy Mountains in the Morning Moon)
Zuni	Moon No Name Moon (same as May)

The twelfth full moon of the year, also the full moon before the winter solstice.

Colonial American	Christmas Moon (also Christ's Moon, Long Night Moon, Moon before Yule, Oak Moon)
Algonquin	Cold Moon (also Long Night's Moon)
Cherokee	Month of the Snow Moon
Cheyenne	Big Freezing Moon
Choctaw	Peach Moon
Dakotah Sioux	Twelfth Moon
Haida	Kungyadikadas (no translation)
Kutenai	Nistamu Natanik (no translation)
Laguna	Middle Winter Moon
Lakota Sioux	Moon of the Popping Trees
Micmac	The Chief Moon
Mohawk	Time of Cold
Natchez	Bears Moon
Nez Perce	Heekui (no translation)
Ojibway	The Little Spirit Moon
Osage	Baby Bear Moon
Oto	Cold Month Moon
Ponca	Beginning of Cold Weather with Snow Moon
San Ildefonso	Ashes Fire Moon
San Juan	Ashes Fire Moon
Taos	Night Moon
Tlingit	Ground-Hog Mother's Moon
Wisham	Her Winter Houses Moon
Zuni	Turning Moon (same as June)

RARE MOONS

A blue moon is popularly defined as a full moon that occurs twice in the same calendar month. When the date for one full moon falls on or near the beginning of a calendar month, the following full moon — always about 29½ days later — comes before the end of the month. February has only 28 days (except for leap years), so there is never a blue moon in this month. Blue moons occur approximately seven times every 19 years, an average of once every 33 months, or 37 per century.

Why are they called "blue moons"? According to research by Philip Hiscock, a professional folklorist at the Memorial University of Newfoundland, the first use of the phrase dates to the 1500s, when it was used to describe something that could never happen. Later, in the mid-1800s, atmospheric pollution from forest fires and volcanic eruptions was linked to a rare, but real event, when the Moon actually appeared to turn a shade of blue.

About the middle of the 1900s, the phrase began to be used to describe the second full moon in a calendar month — another "rare," but real event — but it didn't catch on until the 1980s and when it did, it was by mistake. The original use referred to a fourth full moon during a three-month calendar season, Spring, Summer, Winter, or Fall. Under this definition, blue moons could only occur in the months just before an equinox or solstice, appearing in the third week of February, May, August, or November. An article in *Sky & Telescope Magazine* in 1946 incorrectly described the event and due to the subsequent widespread usage, the meaning has long been over-shadowed by the current application.

Blue moons have no real significance but arouse interest because they don't appear very often. Intriguingly, those months that do have blue moons are the same periods recognized as "leap months" in two older calendar cultures. The traditional Chinese and Hindu calendars are based on a lunisolar system, balancing the cycles of both the Moon and the Sun. In order not to get out of step, these calendars must periodically adjust dates and the adjustment periods are the same months in which there are blue moons.

Blue moons may sometimes cause confusion because of differences in local time zones. A full moon that falls on the first day of the month but only a few hours after midnight, for example, will produce a second full moon in that month. But a few time zones to the west, the local time of the first full moon will actually fall on the last day of the preceding month, making the blue moon a month earlier.

Less noticeable than blue moons but equally rare are months where there are two new moons, but the rarest calendar phenomenon of all is a month when there are no full moons. This can only happen in February and the event only occurred four times in the 1900s, the last in February 1999. In the twenty-first century, this kind of full moonless month will also occur four times, in 2018, 2037, 2067, and 2094. On each of these occasions, the month before and the month after will both have blue moons, because one event cannot happen without the other.

Even though a blue moon as we now know it is not colored blue, unusual atmospheric conditions can change lunar color. In some

> ## BLUE MOON VITAL STATISTICS
>
> On average, blue moons occur:
>
> Once every 2.7 years
>
> 7 times every 19 years
>
> Once every 33 months
>
> 37 times every century
>
> Once every 33 full moons
>
> Some of these figures were adapted from an article published in *The Planetarian*, (12/1993)

cases, the Moon may appear to be tinted blue. A blue-colored moon or one with a greenish color is most likely seen just before sunrise or just after sunset if there is a large quantity of dust or smoke particles high in the sky. These particles can filter out the colors of longer wavelengths such as red and yellow, leaving only green and blue. In 1883, after the Krakatoa volcano in Indonesia erupted, observers around the world noted that wildly colored sunsets were common for most of the following two years. At the same time and for the same reason, the Moon appeared blue.

MOON OF MANY FACES

In some western cultures, popular legends and mythology describe a "man in the moon." This familiar image comes from the unique markings seen on the face of the moon, light and dark areas on the surface that seem to form the shape of a human face. In some cultures, different images characterize the surface patterns, including a "lady in the moon" and a "rabbit in the moon." Other animals with claims to the lunar image include a beetle, a toad, a fox, a cow, a cat, a bear, and a lion. Sometimes, the patterns on the surface suggest different images at different stages in the Moon's path across the sky, because the arc of the path "tilts" the Moon relative to observers on Earth.

LADY IN THE MOON

RABBIT IN THE MOON

BEETLE IN THE MOON

MAN IN THE MOON

THE VITAL STATISTICS OF THE MOON

MEAN DISTANCE OF THE MOON FROM EARTH	238,712 miles (384,400 km) 60.27 Earth radii 0.002 570 a.u.
GREATEST DISTANCE OF THE MOON FROM EARTH (APOGEE)	252,586 miles (406,740 km)
SHORTEST DISTANCE OF THE MOON FROM EARTH (PERIGEE)	221,331 miles (356,410 km)
CIRCUMFERENCE	6,790 miles (10,930 km) 0.27 of Earth's circumference
DIAMETER	2,160 miles (3,476 km) 0.27 of Earth's diameter
MEAN RADIUS	1,079 miles (1,737.5 km)
EQUATORIAL RADIUS	1,079 miles (1,738 km)
POLAR RADIUS	1,077 miles (1,735 km)
MEAN ANGULAR DIAMETER	31' 07"

ABBREVIATIONS

See GLOSSARY (page 134) for explanations of unfamiliar terms.

km	kilometer	**g**	gravity
kg	kilogram	**m**	meter
a.u.	astronomical unit	**s or sec**	second
cm	centimeter	**hr**	hour

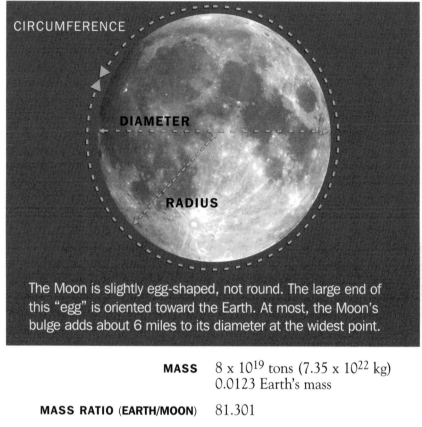

CIRCUMFERENCE

DIAMETER

RADIUS

The Moon is slightly egg-shaped, not round. The large end of this "egg" is oriented toward the Earth. At most, the Moon's bulge adds about 6 miles to its diameter at the widest point.

MASS	8×10^{19} tons (7.35×10^{22} kg) 0.0123 Earth's mass
MASS RATIO (EARTH/MOON)	81.301
VOLUME	2.4×10^{9} miles^3 (2.197×10^{10} km^3) 0.0203 Earth's volume
MEAN DENSITY	208 lb/ft^3 (3.34 g/cm^3) 3.33 more dense than water 0.6 Earth's density
GRAVITY AT THE SURFACE	5.31 ft/sec^2 (1.62 m/s^2) 0.1667 g (1/6 Earth's gravity)
ESCAPE VELOCITY	1.48 miles/sec (2.38 km/sec)
MEAN INCLINATION TO LUNAR EQUATOR	6° 41'

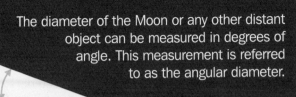

The diameter of the Moon or any other distant object can be measured in degrees of angle. This measurement is referred to as the angular diameter.

Depending on where it is in its elliptical orbit, the angular diameter of the Moon varies from 29' 33" to 33' 6", only about half a degree. By comparison, at arm's length the human fist measures 10 degrees and a single finger represents about 2 degrees.

MEAN ORBITAL INCLINATION TO ECLIPTIC	5° 08' 43"
OSCILLATION OF ORBITAL INCLINATION TO EQUATOR	± 0° 9' every 173 days
INCLINATION OF LUNAR EQUATOR TO ECLIPTIC	1° 32' 33"
PERIOD OF REVOLUTION OF PERIGEE	3,232 days
ORBITAL DIRECTION	east (counterclockwise)
MEAN ORBITAL SPEED	2,287 miles/hour (3,683 km/hr) 33 minutes arc/hour
DAILY SIDEREAL MOTION	13.176358 degrees
MEAN CENTRIPETAL ACCELERATION	0.01 ft/s^2 (0.00272 m/s^2) 0.0003 g

MEAN ECCENTRICITY OF ORBIT	0.0549 (mean eccentricity of Earth's orbit is 0.0167)
SYNODIC MONTH (NEW MOON TO NEW MOON)	29.53059 days 29 days, 12 hr, 44 min, 2.8 sec
SIDEREAL MONTH (STAR TO STAR)	27.32166 days 27 days, 7 hr, 43 min, 11.5 sec
ANOMALISTIC MONTH (APOGEE TO APOGEE OR PERIGEE TO PERIGEE)	27.55455 days 27 days, 13 hr, 18 min, 33.2 sec
NODICAL MONTH, DRACONIC MONTH (NODE TO NODE)	27.21222 days 27 days, 5 hr, 5 min, 35.8 sec
TROPICAL MONTH (FIRST POINT OF ARIES TO FIRST POINT OF ARIES)	27.321582 days 7 days, 7 hr, 43 min, 4.7 sec
REGRESSION OF NODES	6,798 days 18.6247 years (19.538 degrees per year)
ROTATION PERIOD	27.321661 days 27 days, 7 hr, 43 min, 11.5 sec
SURFACE TEMPERATURE	273° F (120° C) day –244° F (–153° C) night
SURFACE AREA	14,657,449 miles^2 (37,958,621 km^2) 9.4 billion acres 26% larger than Africa
VISIBLE SURFACE	41 percent during one lunar cycle 18 percent additional surface visible due to librations 59 percent total visible surface
PARALLAX	0.9507 degrees

The Moon's orbital distance
at apogee is equivalent
to 32 Earth diameters

The Moon's orbital distance
at perigee is equivalent
to 28 Earth diameters

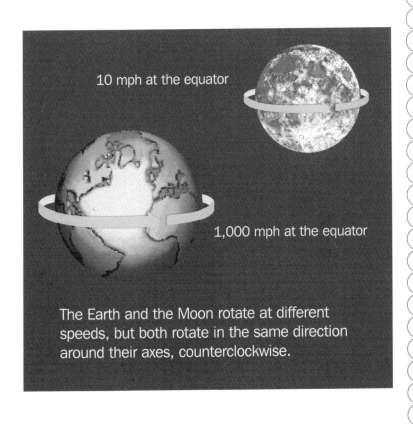

10 mph at the equator

1,000 mph at the equator

The Earth and the Moon rotate at different
speeds, but both rotate in the same direction
around their axes, counterclockwise.

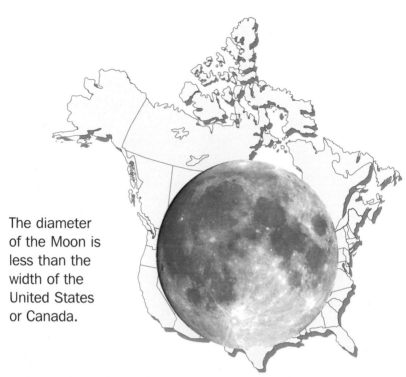

The diameter of the Moon is less than the width of the United States or Canada.

The Earth and the Moon at the same scale. The Moon is a little more than one quarter the diameter of the Earth.

Earth photograph courtesy of NASA;
Moon photograph courtesy of Lick Observatory.

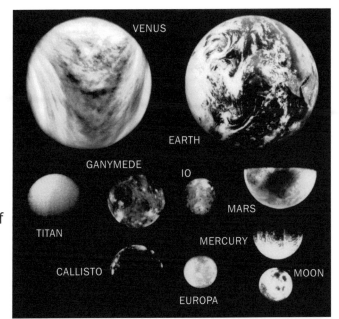

A size comparison of the Moon with some of the planets of the solar system and the moons of Jupiter.

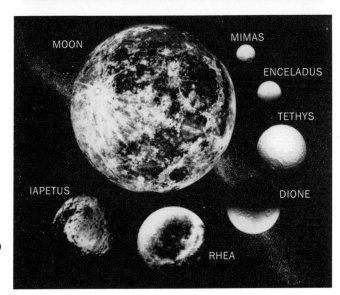

The Moon compared to the moons of Saturn.

Photographs courtesy of NASA

MOON'S ANGULAR DIAMETER	0.5181 degrees
MAGNITUDE OF FULL MOON	−12.5
AVERAGE ALBEDO	0.07
ESTIMATED AGE OF MOON	4.6 billion years
FLIGHT TIME FROM EARTH	60 to 70 hours
INCREASE IN MEAN DISTANCE FROM EARTH	1½ inches/year (3.8 cm/year)

TIME CONVERSION CHART

Universal Time (UT) is also known as Greenwich Mean Time.

UT	AST Atlantic Standard Time	ADT Atlantic Daylight Time	EST Eastern Standard Time	EDT Eastern Daylight Time	CST Central Standard Time	CDT Central Daylight Time	MST Mountain Standard Time	MDT Mountain Daylight Time	PST Pacific Standard Time	PDT Pacific Daylight Time
00	8PM	9PM	7PM	8PM	6PM	7PM	5PM	6PM	4PM	5PM
01	9	10	8	9	7	8	6	7	5	6
02	10	11	9	10	8	9	7	8	6	7
03	11	MIDNIGHT	10	11	9	10	8	9	7	8
04	MIDNIGHT	1AM	11	MIDNIGHT	10	11	9	10	8	9
05	1AM	2	MIDNIGHT	1AM	11	MIDNIGHT	10	11	9	10
06	2	3	1AM	2	MIDNIGHT	1AM	11	MIDNIGHT	10	11
07	3	4	2	3	1AM	2	MIDNIGHT	1AM	11	MIDNIGHT
08	4	5	3	4	2	3	1AM	2	MIDNIGHT	1AM
09	5	6	4	5	3	4	2	3	1AM	2
10	6	7	5	6	4	5	3	4	2	3
11	7	8	6	7	5	6	4	5	3	4
12	8	9	7	8	6	7	5	6	4	5
13	9	10	8	9	7	8	6	7	5	6
14	10	11	9	10	8	9	7	8	6	7
15	11	NOON	10	11	9	10	8	9	7	8
16	NOON	1PM	11	NOON	10	11	9	10	8	9
17	1PM	2	NOON	1PM	11	NOON	10	11	9	10
18	2	3	1PM	2	NOON	1PM	11	NOON	10	11
19	3	4	2	3	1PM	2	NOON	1PM	11	NOON
20	4	5	3	4	2	3	1PM	2	NOON	1PM
21	5	6	4	5	3	4	2	3	1PM	2
22	6	7	5	6	4	5	3	4	2	3
23	7	8	6	7	5	6	4	5	3	4

WEIGHT ON THE MOON

EARTH WEIGHT	MOON WEIGHT	EARTH WEIGHT	MOON WEIGHT	EARTH WEIGHT	MOON WEIGHT
50 lbs	8.50 lbs	85	14.45	120	20.40
51	8.67	86	14.62	121	20.57
52	8.84	87	14.79	122	20.74
53	9.01	88	14.96	123	20.91
54	9.18	89	15.13	124	21.08
55	9.35	90	15.30	125	21.25
56	9.52	91	15.47	126	21.42
57	9.69	92	15.64	127	21.59
58	9.86	93	15.81	128	21.76
59	10.03	94	15.98	129	21.93
60	10.20	95	16.15	130	22.10
61	10.37	96	16.32	131	22.27
62	10.54	97	16.49	132	22.44
63	10.71	98	16.66	133	22.61
64	10.88	99	16.83	134	22.78
65	11.05	100	17.00	135	22.95
66	11.22	101	17.17	136	23.12
67	11.39	102	17.34	137	23.29
68	11.56	103	17.51	138	23.46
69	11.73	104	17.68	139	23.63
70	11.90	105	17.85	140	23.80
71	12.07	106	18.02	141	23.97
72	12.24	107	18.19	142	24.14
73	12.41	108	18.36	143	24.31
74	12.58	109	18.53	144	24.48
75	12.75	110	18.70	145	24.65
76	12.92	111	18.87	146	24.82
77	13.09	112	19.04	147	24.99
78	13.26	113	19.21	148	25.16
79	13.43	114	19.38	149	25.33
80	13.60	115	19.55	150	25.50
81	13.77	116	19.72	151	25.67
82	13.94	117	19.89	152	25.84
83	14.11	118	20.06	153	26.01
84	14.28	119	20.23	154	26.18

EARTH WEIGHT	MOON WEIGHT	EARTH WEIGHT	MOON WEIGHT	EARTH WEIGHT	MOON WEIGHT
155	26.35	193	32.81	54	9.18
156	26.52	194	32.98	55	9.35
157	26.69	195	32.15	56	9.52
158	26.86	196	33.32	57	9.69
159	27.03	197	33.49	58	9.86
160	27.20	198	33.66	59	10.03
161	27.37	199	33.83	60	10.20
162	27.54	200	34.00	61	10.37
163	27.71			62	10.54
164	27.88	25 kg	4.25 kg	63	10.71
165	28.05	26	4.42	64	10.88
166	28.22	27	4.59	65	11.05
167	28.39	28	4.76	66	11.22
168	28.56	29	4.93	67	11.39
169	28.73	30	5.10	68	11.56
170	28.90	31	5.27	69	11.73
171	29.07	32	5.44	70	11.90
172	29.24	33	5.61	71	12.07
173	29.41	34	5.78	72	12.24
174	29.58	35	5.95	73	12.41
175	29.75	36	6.12	74	12.58
176	29.92	37	6.29	75	12.75
177	30.09	38	6.46	76	12.92
178	30.26	39	6.63	77	13.09
179	30.43	40	6.80	78	13.26
180	30.60	41	6.97	79	13.43
181	30.77	42	7.14	80	13.60
182	30.94	43	7.31	81	13.77
183	31.11	44	7.48	82	13.94
184	31.28	45	7.65	83	14.11
185	31.45	46	7.82	84	14.28
186	31.62	47	7.99	85	14.45
187	31.79	48	8.16	86	14.62
188	31.96	49	8.33	87	14.79
189	32.13	50	8.50	88	14.96
190	32.30	51	8.67	89	15.13
191	32.47	52	8.84	90	15.30
192	32.64	53	9.01		

RESOURCES

This book, even if it were twice as large, could not contain more than a fraction of the information that is known about the moon. For those fans of the moon who want to learn more, there are numerous resources available. The following books are suggested for further reading, but not all of these titles may be in print or available for sale at bookstores. To locate a particular title, try your local public or university library, the publisher, or an online book retailer. Also, see the *Bibliography* for additional useful resources.

The Astronomical Companion, by Guy Ottewell. 1979, Universal Workshop at Furman University (Greenville, SC).

Astronomy, A Step-by-Step Guide to the Night Sky, by Storm Dunlop. 1985, Macmillan Publishing Company.

Eclipse!: the What, Where, When, Why & How Guide to Watching Solar & Lunar Eclipses, by Philip S. Harrington. 1997, John Wiley & Sons.

Exploring the Moon through Binoculars and Small Telescopes, by Ernest H. Cherrington, Jr. 1984, Dover, Publications.

The Geology of the Terrestrial Planets, by Carr, Saunders, Stron, and Wilhelms. 1984, NASA. (NASA # SP-469).

Luna: Myth & Mystery, by Kathleen Cain. 1991, Johnson Books.

The Moon, by Patrick Moore. 1981, Rand McNally & Company.

The Moon Observer's Handbook, by Fred W. Price. 1988, Cambridge University Press.

Moon Shot: The Inside Story of America's Race to the Moon, by Alan Shepard and Deke Slayton. 1994, Turner Publishing.

The Once and Future Moon, by Paul D. Spudis. 1996, Smithsonian Institution Press.

Pictorial Guide to the Moon, by Dinsmore Alter. 1979, Crowell.

Welcome to Moonbase, by Ben Bova. 1987, Ballantine Books.

CALENDARS AND ANNUALS

The Astronomical Calendar. Created annually by Guy Ottewell, the author of the *Astronomical Companion* (see above). This large format publication is a detailed guide to the astronomical events for each month. Clear, well-illustrated descriptions of the major astronomical cycles, including those of the Moon. Also includes a libration guide, graphic depictions of eclipses, and an illustrated guide for sighting the first crescent moon. Available in bookstores, telescope stores, and directly from the publisher:

Universal Workshop at Furman University, Greenville, S.C. 29613. 864-294-2208 http://www.kalend.com

Astronomical Phenomena. Usually available 1–2 years in advance of a calendar year. The major resource for listings of specific solar and lunar cycles, including phases, settings, and risings. Published by the U.S. Naval Observatory. For sale from U.S. Government Bookstores or from the U.S. Government Printing Office. Some contents are also available online. http://www.usno.navy.mil/

Harris' Farmer's Almanac. Issued annually. A traditional almanac format, with daily information about astronomical events, including rising and setting times. Available in bookstores and newsstands. Harris' Farmer's Almanac, P.O. Box 658, Sharon, CT 06069

The Moon Calendar. Issued annually beginning in August preceding the calendar year. Large black and white poster with accurate images of the moon's phase for each day of the year. Available at local bookstores, museum stores, or telescope stores. Johnson Books, 1880 South 57th Court, Boulder, CO 80301   303-443-1576

Observer's Handbook. Issued annually in the fall preceding the calendar year. A detailed collection of astronomical events, including moon cycles, rising, setting, occultations, librations, and eclipses. Limited distribution in the U.S. Royal Astronomical Society of Canada, 136 Dupont Street, Toronto, Ontario M5R 1V2 Canada   416-924-7973   http://www.rasc.ca

The Old Farmer's Almanac. Issued annually (usually in September preceding the calendar year) by Yankee Publishing. Includes moonrise and moonset times as well as other basic astronomical data. For sale at most newsstands and bookstores. Yankee Publishing, Main Street, Dublin, NH 03444   http://www.almanac.com/

PERIODICALS

Astronomy Magazine. Kalmbach Publishing Company, P.O. Box 1612, Waukesha, WI 53187   414-796-8776 http://www.kalmbach.com/astro/astronomy.htm/

Griffith Observer. Griffith Observatory, 2800 East Observatory Road, Los Angeles, CA 90027   213-664-1181   http://www.GriffithObs.org/

The Planetarian. International Planetarian Society, c/o Griffith Observatory, 2800 East Observatory Road, Los Angeles, CA 90027   213-664-1181 http://www.GriffithObs.org/IPSPlanetarian.html/

Sky and Telescope Magazine. Sky Publishing Corporation. P.O. Box 9111, Belmont, MA 02178-9111   617-864-7360   http://www.skypub.com/

Sky News: The Canadian Magazine of Astronomy & Stargazing. National Museum of Science and Technology, P.O. Box 9724, Station T, Ottawa, Ontartio K1G 5A3 800-267-3999

The Strolling Astronomer. Official publication of the Association of Lunar and Planetary Observers. ALPO, P.O. Box 171302, Memphis, TN 38187 http://www.lpl.arizona.edu/alpo

ORGANIZATIONS

American Astronomical Society. 2000 Florida Avenue NW Unit 400, Washington, DC 20009 202-328-2010 http://www.aas.org/

Association of Lunar and Planetary Observers (ALPO). Box 3AZ, University Park, NM 88003 http://www.lpl.arizona.edu/alpo/

Astronomical League. Science Service Bldg, 1719 N St. NW, Washington, DC 847-398-0562 http://www.mcs.net/~bstevens/al/index.html/

Astronomical Society of the Pacific. 390 Ashton Ave., San Francisco, CA 94112 415-337-1100 http://www.aspsky.org/

Lunar & Planetary Institute. 3600 Bay Area Blvd., Houston, TX 77058 281-486-2139 http://cass.jsc.nasa.gov/

National Space Society. 600 Pennsylvania Ave. SE Unite 201, Washington, DC 20003 202-543-1900 http://www.nss.org/

Royal Astronomical Society of Canada. 136 Dupont Street, Toronto, Ontario M5R 1V2 Canada 416-924-7973 http://www.rasc.ca

MAPS

The Earth's Moon. This large color map has labeled details of both front and back sides of the Moon. Published by the National Geographic Society, and distributed by Geosystems, 1350 Pine Street, Boulder, CO 80302, 888-494-6277

Lunar Chart (LPC-1). The Defense Mapping Agency Aerospace center created this map for NASA. In a large color format, the map includes a detailed mercator projection of the surface between 45° north and 45° south, as well as stereographic projections of both polar regions. Available from the National Space Science Data Center, Goddard Space Flight Center, Greenbelt, MD 20771.

Moon Map. The publisher of *Sky & Telescope Magazine* has also created this small (11 inch diameter) map of major features of the lunar surface. Published by Sky Publishing Corporation, P.O. Box 9111, Belmont, MA 02178-9111 617-864-7360 http://www.skypub.com

Official Map of the Moon. This large poster-sized color map of the lunar surface includes detailed labeling of features. Published by Rand McNally, 8255 N. Central Park, Skokie, IL 60076 847-329-8100

Other maps and surface imagery. The U.S. Geological Survey is the official agency responsible for maps of the lunar surface. General information: USGS Information Services, Denver, CO 303-202-4700 http://wwwflag.wr.usgs.gov/USGSFlag/Space/wall/moon.html/

SOFTWARE

Software that has been created about astronomy, the stars, or the solar system often includes information, images, and interactive features about the Moon and its cycles. Planetarium programs also frequently include the Moon as part of their features. New programs may be discovered from reviews and announcements in astronomy publications such as *Sky & Telescope* and *Astronomy.* Check with the publisher for information about upgrades.

Astronomy Lab (PC). Personal MicroCosms, 8547 East Arapahoe Road, Unit J-147, Greenwood Village, CO 80112 303-753-3268 http://www.users.aol.com/eric98398/index.htm/

Distant Suns (PC, Mac). Andromeda Software Inc., P.O. Box 605-N, Amherst, NY 14226-0605 http://www.frii.com/~astron/index.htm/

Deep-Sky Planner (PC, Mac). Sky Publishing Company, P.O. Box 9111, Belmont, MA 02178 617-864-7360 http://www.skypub.com/catalog/dsp20/dsplan.htm/

Interactive Astronomical Almanac (PC, Mac). Product code: 8440. National Technical Information Service, 5285 Port Royal Road, Springfield, VA 22161 703-605-6000

Voyager (Mac). Carina Software, 12919 Alcosta Blvd., Unit 7, San Ramon, CA 94583 925-355-1266 http://www.carinasoft.com/

Moon Calculator (PC). http://www.starlight.demon.co.uk/mooncalc/

Moon Clock (Mac). SoftTouch Applications, 7742 E. Oakwood Circle, Tucson, AZ 85750 http://rtd.com/~gschneid/softtouch.html/

Moonrise (PC). http://www.iserv.net/~bsidell/moonrise.htm/

Moontimes (PC, MAC). Zephyr Services, 1900 Murray Avenue, Pittsburgh, PA 15217 412-422-6600 http://www.zephyrs.com/

MoonTimes (PC). Totality Software Inc., 9974 Scripps Ranch Road Blvd, Unit 305, San Diego, CA 92131 619-234-8600

Multiyear Interactive Computer Almanac (PC). Willmann-Bell, Inc., Box 35025, Richmond, VA 23235 804-320-7016 http://www.willbell.com/

New Moons (PC). Zephyr Services, 1900 Murray Avenue, Pittsburgh, PA
15217 412-422-6600 http://www.zephyrs.com/

Sky Calc (PC). Software Bisque, 912 12th Street, Unit A, Golden, CO
80401 303-278-4478 http://www.bisque.com/

Skyclock. (PC). Dynacomp, Inc., 4560 East Lake Road, Livonia, NY 14487
716-346-9788

CALCULATIONS

Astronomical Formulae for Calculators, 4th Edition, by Jean Meeus. 1988,
Willmann-Bell, Inc., Box 35025, Richmond, VA 23235 804-320-7016

Astronomical Tables of the Sun, Moon and Planets, Second Edition, by Jean
Meeus. 1995, Willmann-Bell, Inc. (see above)

*Explanatory Supplement to the Astronomical Ephemeris and the American
Ephemeris and Nautical Almanac.* 1961, by the U.S. Naval Observatory.

Practical Astronomy with Your Calculator, Second Edition, by Peter Duffett-
Smith. 1981, Cambridge University Press.

ONLINE

The contents and organization of online services evolve rapidly. Some of the
products and services listed may change names or sites or new resources may
become available at any time. For the best results, search with one or more
appropriate keywords, including: moon, lunar, astronomy, eclipse, moonrise,
phase, etc.

Adler Planetarium http://astro.uchicago.edu/adler/

American Astronomical Society http://www.aas.org/

Association of Lunar & Planetary Observers
http://www.jpl.arizona.edu/alpo/

Astronomical League http://www.mcs.net/~bstevens/al/

Astronomical Society of the Pacific http://www.aspsky.org/

Astronomy Magazine http://www.kalmbach.com/astro/astronomy.htm/

Griffith Observatory http://www.GriffithObs.org/

International Occultation Timing Association
http://www.sky.net/~robinson/iotandx.htm/

Jet Propulsion Laboratory http://www.jpl.nasa.gov/

Lunar & Planetary Society http://cass.jsc.nasa.gov/

NASA http://www.nasa.gov/

National Space Science Center http://nssdc.gsfc.nasa.gov/

North American Skies http://www.webcom.com/safezone/NAS/

Sky & Telescope Magazine http://www.skypub.com/

Solar and Lunar Eclipse information
http://planets.gsfc.nasa.gov/eclipse/eclipse.html/

U.S. Naval Observatory http://www.usno.navy.mil/

TELESCOPES, CCDS, ASTRONOMICAL EQUIPMENT

Apogee, 3340 N. Country Club, Tucson, AZ 85716 520-326-3600
http://www.apogee-ccd.com/

Astro Link, P.O. Box 1978, Spring Valley, CA 92077 619-449-4722

Celestron Telescopes, P.O. Box 3578, Torrance, CA 90503 800-421-1526
http://www.celestron.com/main.htm/

Edmund Scientific Company, Edscorp Building, Barrington, NJ 08007 609-573-6250 http://www.edsci.com/

Meade Instruments Corporation, 1675 Toronto Way, Costa Mesa, CA 92626 714-556-2291 http://www.meade.com/

Orion Telescopes, P.O. Box 1815, Santa Cruz, CA 95061 408-763-7000

S & S Optika, 5172 S. Broadway, Englewood, CO 80110 303-789-1089

Santa Barbara Instrument Group (SBIG), P.O. Box 50437, Santa Barbara, CA 93150 805-969-1851 http://www.sbig.com

Software Bisque, 912 12th Street, Unit A, Golden, CO 80401 303-278-4478 http://www.bisque.com

SpectraSource, 31324 Via Colinas, Unit 114, Westlake Village, CA 91362 818-707-2655

BIBLIOGRAPHY

Alter, Dinsmore. *Pictorial Guide to the Moon, Third Revised Edition*. 1973, Thomas Y. Crowell Company (New York).

Bredon, Juliet, and Mitrophanow, Igor. *The Moon Year: A Record of Chinese Customs and Festivals*. 1927, Kelly & Walsh, Ltd. (Shanghai).

British Astronomical Association. *Guide to Observing the Moon*. 1986, Enslow Publishers, Inc. (Hillside, NJ).

Cadogan, Peter H. *The Moon: Our Sister Planet*. 1981, Cambridge University Press.

Carr, Michael H., editor. *The Geology of the Terrestrial Planets*. 1984, National Aeronautics and Space Administration.

Corliss, William R. *The Moon and the Planets: A Catalog of Astronomical Anomalies*. 1985, The Sourcebook Project (Glen Arm, Maryland).

Davidson, Norman. *Sky Phenomena: A Guide to Naked-Eye Observation of the Stars*. 1993, Lindisfarne Press (Hudson, NY).

Dershowitz, Nachum, and Reingold, Edward M. *Calendrical Calculations*. 1997, Cambridge University Press.

Duffet-Smith, Peter. *Practical Astronomy with Your Calculator, Second Edition*. 1981, Cambridge University Press.

Dunlop, Storm. *Astronomy, A Step-by-Step Guide to the Night Sky*. 1985, Macmillan Publishing Company.

Firsoff, V.A. *Strange World of the Moon*. 1959, Basic Books.

French, Bevan M. *The Moon Book*. 1977, Penguin Books.

Harrington, Philip S. *Eclipse!: The What, Where, When, Why & How Guide to Watching Solar & Lunar Eclipses*. 1997, John Wiley & Sons, Inc.

Heiken, Grant, Vaniman, David, and French, Bevan M. *Lunar Sourcebook: A User's Guide to the Moon*. 1991, Cambridge University Press.

Kaler, James B. *The Ever-Changing Sky: A Guide to the Celestial Sphere*. 1996, Cambridge University Press.

Katzeff, Paul. *Full Moons: Fact and Fantasy about Lunar Influence*. 1981, Citadel Press/Lyle Stuart Inc.

Kelly, Adrian, Dresser, Peter, and Ross, Linda M. *Religious Holidays and Calendars*. 1993, Omnigraphics (Detroit, MI).

Martinez, Patrick, editor. *The Observer's Guide to Astronomy*. 1994, Cambridge University Press.

Meeus, Jean. *Astronomical Algorhythms*. 1991, Willman-Bell, Inc. (Richmond, VA).

Meeus, Jean. *Astronomical Formulae for Calculators, Fourth Edition*. 1988, Willman-Bell, Inc. (Richmond, VA).

Meeus, Jean. *Astronomical Tables of the Sun, Moon, and Planets, Second Edition*. 1995, Willman-Bell, Inc. (Richmond, VA).

Menzel, Donald H. *A Field Guide to the Stars and Planets*. 1964, Peterson Field Guide Series/Houghton Mifflin Company.

Moore, Patrick. *The Moon*. 1981, Mitchell Beazley Publishers/Rand McNally & Company.

Muirden, James, editor. *Sky Watcher's Handbook*. 1993, W.H. Freeman and Company Limited.

Ottewell, Guy. *The Astronomical Companion*. 1979, Universal Workshop at Furman University (Greenville, SC).

Parise, Frank. *The Book of Calendars*. 1982, Facts On File, Inc.

Price, Fred W. *The Moon Observer's Handbook*. 1988, Cambridge University Press.

Rackham, Thomas. *Moon in Focus*. 1968, Pergamon Press Ltd. (London, U.K.).

Rükl Antonín. *Atlas of the Moon*. 1990, Kalmbach Books (Waukesha, WI).

Seidelmann, P. Kenneth, editor. *Explanatory Supplement to The Astronomical Almanac*. 1992, University Science Books (Mill Valley, CA).

Spudis, Paul D. *The Once and Future Moon*. 1996, Smithsonian Institution Press.

Westrheim, Margo. *Calendars of the World: A Look at Calendars and the Ways We Celebrate*. 1993, Oneworld Publications (Oxford, England).

Whipple, Fred L. *Earth, Moon, and Planets, Third Edition*. 1970, Harvard University Press.

Wylie, Francis E. *Tides and the Pull of the Moon*. 1979, Berkley Publishing Corporation.

GLOSSARY

albedo   The percentage of light reflected from the surface of a planet or Moon. Albedo is determined by measuring the ratio between the light reflected and the light shining on an object; complete reflection is represented by an albedo of 1.

angular diameter   The diameter of a distant object as measured by the angle formed from a point representing an observer and the outer edges of the object.

annular eclipse   An eclipse of the Sun when the Moon is farthest away in its orbit around Earth. At this point, its apparent diameter is not large enough to completely obscure the sun. During an annular eclipse, a ring of light is left uncovered around the dark circle produced by the Moon.

anomalistic month   The period of time it takes the Moon to go from one point of apogee (or perigee) to the next: 27.55455 days.

anorthositic rock   One of the types of rocks found on the Moon at higher elevations.

aphelion   The point in a planet's orbit around the Sun when it is farthest from the Sun (opposite of perihelion).

apogee   The point in the Moon's orbit when it is farthest from the Earth (opposite of perigee).

apogean tide   The low tide of the month that occurs when the Moon is at apogee (farthest from Earth).

apolune   The point in the orbit of an object around the Moon (such as a spacecraft) when it is farthest from the Moon's surface.

ASA   The code used to designate the light sensitivity of photographic film. These codes are designed so that the higher numbers represent greater sensitivity and lower numbers less sensitivity.

asteroid   A body of rock or frozen liquid that is in orbit around the Sun. Asteroids are sometimes considered planetoids, or small planets.

astronomical unit (A.U.)   The mean distance between the Sun and the Earth and used as a standard of measurement. 1 A.U. = 92,955,630 miles (149,597,870 km).

astronomy	The science dealing with objects in space.
astrophysics	A branch of astronomy, using physics to study and explain celestial objects.
axis	An imaginary line through the center of mass of an object, around which the object rotates.
azimuth	The angle measured from due north of an observer to directly under an object of interest in the sky. With the observer facing south, north is 0 degrees, east is 90 degrees, south is 180 degrees, and west is 270 degrees.
barycenter	A point marking the center of mass created when two celestial objects orbit around each other.
basalt	A type of igneous rock created from lava and found on the lunar surface in low areas, also found on Earth.
breccia	A composite rock found on the Moon and formed from small pieces of different minerals, also found on Earth.
CCD	Charge-coupled device. A camera or the component of a camera that registers and records images.
celestial equator	An imaginary extension of the Earth's equator into the sky. The celestial equator is 90 degrees from each of Earth's celestial poles.
celestial mechanics	The branch of astronomy dealing with the motions and gravitational effects of celestial objects.
circumference	The linear measurement around the outside of a circle or a sphere.
colongitude	The longitude on the surface of the Moon marked by the terminator, the edge of the area illuminated by the Sun.
conjunction	The position of two celestial bodies when they are in line with one another as seen by an observer on Earth. The new moon is also referred to as moon in conjunction with the Sun (opposite of opposition).
crater wall	The circular wall formed by the impact of a meteorite on the lunar surface.
crescent moon	A phase of the Moon just before and after the new moon, when only a thin curved section is lighted by the Sun. The last crescent moon before the new moon is sometimes called the old crescent moon, and the first crescent moon

135

after the new moon is sometimes called the young crescent moon.

culmination The highest point a celestial body reaches in the sky as seen from Earth, always occurring when the body's azimuth is 180, or due south.

dark of the moon Another name for the new moon.

Daylight Saving Time (DST) A legislated time change in some countries in which local times are moved up by one hour in the spring and back one hour in the fall ("spring ahead, fall back").

declination The angle measured between the celestial equator and an object in the sky.

density An object's mass divided by its volume (grams/centimeter).

diurnal Referring to a period of one day.

DST See Daylight Saving Time.

earthshine Reflected light from the Earth, visible as a dull, red, or copper glow on the Moon during lunar eclipses. Earth-shine can also sometimes illuminate a young crescent moon so that the whole face of the Moon can be faintly seen. This effect is often referred to as the "old moon in the new moon's arms."

eclipse The blocking of light from the Sun when the Earth comes between the Sun and the Moon or the Moon between the Sun and the Earth.

ecliptic The imaginary plane formed by the Earth's orbit around the Sun or the plane formed by the apparent motion of the Sun through the sky.

elliptical orbit A non-circular path formed when a body moves around another. The shape is that of a "stretched" or distorted circle.

elongation The angle of a planet away from the Sun or the Moon from the Earth as viewed from the Earth.

ephemeris A publication or list that has information needed to locate a star, moon, or planet in the sky at a particular time.

equatorial tide A tide produced semi-monthly by the position of the Moon over the equator.

escape velocity The speed required for an object to overcome the gravitational force of an astronomical object.

far side   The side of the Moon facing away from the Earth.

first quarter moon   The phase of the Moon when it is 90 degrees away from a line between the Sun and the Earth, measured eastward from the Sun, as seen from the north. The angle of illumination creates a half-circle picture of the Moon's surface, with the lighted half being on the right side.

full moon   The phase of the Moon when it is on the opposite side of the Earth from the Sun and receives sunlight across its entire face, forming a circle of light. At this point, the Moon is in *opposition* to the Sun.

gibbous moon   The phase of the Moon when it is getting larger after the first quarter moon phase (waxing gibbous) or smaller after the full moon but before the last quarter moon (waning gibbous).

gravity   One of the fundamental forces of nature, defined as the constant force of attraction between all objects in the universe. The gravitational force is inversely proportional to the square of the distance between the objects and proportional to the masses.

grazing occultation   An occultation by the Moon of a planet or star where the path of the planet or star only intercepts the north or south limb of the Moon.

Greenwich Mean Time   Time as measured from the 0 degrees longitude position of the Greenwich Observatory in England, also known as Universal Time (UT).

half moon   The phase of the Moon also known as the quarter moon, first quarter moon, or last quarter moon.

intercalation   A method of synchronizing a lunar calendar with a solar year by adding extra days or months. Extra days are known as intercalary days and extra months are known as intercalary months.

lacus   Latin for lake. An area on the surface of the Moon resembling a lake.

last quarter moon   The phase of the Moon when it is 90 degrees away from a line between the Sun and the Earth, measured westward from the Sun, as seen from the north. The angle of illumination creates a half circle of the Moon's surface, with the lighted half being on the left side. Also referred to as the third quarter moon.

latitude Lines of measurement around a planet or the Moon, parallel to its equator. Measured in degrees, with the equator being 0 degrees and the poles 90 degrees north or south.

librations The irregular motions of the Moon in its elliptical orbit around Earth that allow slightly more than half of the Moon's surface to be visible over a period of time.

limb The visible edge of a planet or moon.

longitude Lines of measurement at right angles to the equator of a planet or the Moon. Measured in degrees of angle from a designated line of 0 degrees. On the Moon, 0 degrees longitude is at the center of the visible face, in the Sinus Medii.

lunar day The period of time between two successive transits of the Moon over the same meridian. The mean lunar day is 24.84 hours (1.035 times the mean solar day). Not the same thing as a day on the Moon, which corresponds to a synodic month.

lunar eclipse An eclipse created by the Earth coming between the Sun and the Moon. Lunar eclipses always happen during the full moon phase.

lunar interval The elapsed time between the transit of the Moon over the Greenwich meridian and a local meridian.

lunar rays Visible streaks on the surface of the Moon which radiate away from some craters.

lunartidal interval The length of time between the transit of the Moon and the following high or low tide.

magnitude A numerical value indicating the brightness of an object in space.

mare Latin for sea (plural: maria). An area on the surface of the Moon (or Mars) that is low, dark, and formed from ancient lava flows.

mascon An area of the Moon's surface formed from dense, thick lunar material and having strong local gravitational effects.

mean A mathematical average of a set of numbers or measurements, with the mean equaling the sum of the numbers divided by the number of units. The mean radius of the Moon, for example, is the average radius figured from multiple measurements.

meridian An imaginary line that passes directly north and south through an observer or specified location on Earth. A plane extended from this line into space passes through the zenith (point above the observer).

meteoroid A small body drifting through space. If a meteoroid is pulled into a planet's or moon's gravitational field, it is called a *meteor*; if the meteor survives a trip through the atmosphere and lands, it is called a *meteorite*. If meteorites are not burned up in the process of entering an atmosphere (or if there is no atmosphere), they may strike the surface and if large enough, create craters.

moon The natural satellite of the Earth or the natural satellite of any planet.

moonrise The point in time when the upper limb of the Moon is even with the Earth's horizon as the Moon rises in the east.

moonset The point in time when the upper limb of the Moon is even with the Earth's horizon as the Moon sets in the west.

nadir An imaginary point directly under an observer on the surface of the Earth, extending through the Earth and into the sky (opposite of zenith).

neap tide The lowest high tide of the lunar month, occurring near the first and last quarter moon phases.

near side The side of the Moon facing the Earth.

new moon The phase of the Moon when it is directly between the Earth and the Sun. Because sunlight is hitting only the far side of the Moon, it appears dark from the Earth. Reflected light from Earth can sometimes make the new moon faintly visible.

nodes The imaginary points at which the orbital path of the Moon or other celestial body crosses the ecliptic.

nodical month A lunar cycle measured by the Moon moving from one node and back again: a period of 27.21222 days.

occultation The movement of one celestial object behind another, such as the occultation of the star Spica by the Moon.

old crescent moon Another name for the thin crescent of the Moon that is still illuminated by the Sun before the Moon goes completely dark at the new moon phase.

opposition A specific point in time when a moon or planet is 180 degrees away from the Sun, on the other side of the Earth. The Moon is full when it is in opposition (opposite of conjunction).

orbital eccentricity The degree to which an elliptical orbit is elongated. Measured by the distance between the foci deivided by the major axis.

palus Latin for swamp. An area on the surface of the Moon that is dark and resembles a swamp.

parrallax The perceived movement of a distant object such as a moon, planet, or star due to the movement of the Earth.

partial eclipse A lunar eclipse in which the Moon only partly enters the dark, umbral shadow of the Earth but is inside the secondary, penumbral shadow. Also refers to a solar eclipse when the Moon does not line up completely between the Earth and Sun and only partly obscures the Sun. This type of eclipse also produces a penumbra as well as an umbra.

penumbra The lighter part of a shadow that is formed by diffused light in an area around the edges of an object.

perigee The point in the Moon's orbit when it is closest to the Earth (opposite of apogee).

perigean tide The high tide of the month that occurs when the Moon is at perigee (closest to to Earth).

perihelion The point in a planet's orbit around the Sun when it is closest to the Sun (opposite of aphelion).

perilune The point in the orbit of an object (such as a spacecraft) around the Moon when it is closest to the Moon's surface.

phases The visible changes that the Moon goes through in every lunar month, caused by the changing angle of illumination from the Sun. There are four specific phases — new moon, first quarter moon, full moon, and last quarter moon — and also non-specific phase names such as waxing moon, waning moon, gibbous moon, and crescent moon.

quadrature The position of the Moon or a planet when it is at right angles to the Sun. The Moon is in first quarter phase when it is in east quadrature to the Sun and last quarter phase when it is in west quadrature.

quarter moon The phase of the Moon that can be either the first quarter moon or the last quarter moon. This phase occurs when the Moon is 90 degrees away from a line between the Sun and the Earth. In the northern hemisphere, the angle of illumination creates a half circle picture of the Moon's surface, with the lighted half being on the right side during first quarter moon and on the left side for last quarter moon.

radius The linear measurement from the center of a sphere to the surface, or half of the diameter.

regression of nodes The backwards movement of the Moon's nodes relative to the direction of orbit.

revolution The movement of one body around another in an orbit. Not to be confused with rotation.

rille A valley or small canyon on the surface of the Moon.

rotation The spinning of a body around its own axis. Not to be confused with revolution.

Saros Cycle A cycle of lunar months lasting 18 years and 11.3 days, the time it takes the Moon, the Earth, and the Sun to return to the same position relative to each other.

satellite An object that is in orbit around another object in space.

selenography The science dealing with the study of the surface of the Moon.

selenology The science dealing with the study of the Moon. From the Greek goddess, Selene.

sidereal month A lunar month measured by a return to a specific position marked by a certain star: a period of 27.32166 days.

sinus Latin for bay. An area on the surface of the Moon resembling the bay of an ocean.

solar eclipse An eclipse caused when the Moon comes directly between the Earth and the Sun, temporarily blocking out Sun's disk in the sky.

spring tide The highest tides in a lunar month, occurring near new and full moons, when the Earth, Sun, and Moon are aligned.

synodic month	A lunar month as measured from the point of one new moon to the next new moon: a period of 29.53059 days.
tektites	Small particles on the Moon's surface made of glasslike material and formed from the impact of meteorites.
terminator	The line formed by the edge of the illuminated portion of the Moon.
tides	The cyclical movement of bodies of water or land on the Earth or the Moon caused by the gravitational pull of the Earth, Moon, and Sun.
transit	The point when the path of the Moon, the Sun, a star, or a planet takes it across the meridian.
tropical month	The time required for the Moon to move from the first point of Aries and back: a period of 27.321582 days.
umbra	The darker core of a shadow, usually cone shaped, and surrounded by a lighter penumbral shadow. Also refers to the darker center of sunspots.
waning moon	The period in the Moon's monthly cycle after the full moon and before the new moon. During this period, the lighted portion of the Moon's surface is decreasing.
waxing moon	The period in the Moon's monthly cycle after the new moon and before the full moon. During this period, the lighted portion of the Moon's surface is increasing.
young crescent moon	Another name for the thin crescent of the Moon that is illuminated by the Sun just after the new moon.
zenith	The imaginary point directly above an observer on Earth (opposite of nadir).

INDEX

"From the slimmest crescent to the full moon to the new, this year's array of phases is elegantly, clearly and usefully charted for you."

— ASTRONOMY BOOK CLUB

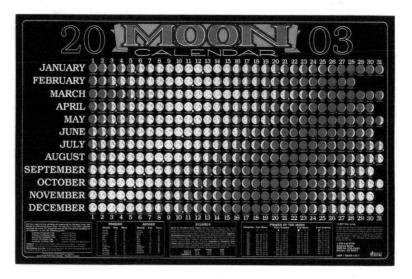

First published in 1981, this accurate annual guide to the Moon's phases is now a standard resource for thousands of amateur and professional astronomers, scientific organizations, museums, planetariums, nature lovers, and Moon watchers.

- Daily illustrations of the phases of the Moon
- Dates of full moons, new moons, quarter moons
- Dates and viewing locations of lunar eclipses
- Times of perigees and apogees
- CARD 10¼" x 6½" and POSTER 31½" x 20½"

Available in local bookstores, museum stores, and other retail outlets, or contact the publisher.

JOHNSON BOOKS 1-800-258-5830